Urban Sustainability

Editor-in-Chief

Ali Cheshmehzangi, Architecture & Built Environment, University of Nottingham Ningbo China, Ningbo, Zhejiang, China

The Urban Sustainability Book Series is a valuable resource for sustainability and urban-related education and research. It offers an inter-disciplinary platform covering all four areas of practice, policy, education, research, and their nexus. The publications in this series are related to critical areas of sustainability, urban studies, planning, and urban geography.

This book series aims to put together cutting-edge research findings linked to the overarching field of urban sustainability. The scope and nature of the topic are broad and interdisciplinary and bring together various associated disciplines from sustainable development, environmental sciences, urbanism, etc. With many advanced research findings in the field, there is a need to put together various discussions and contributions on specific sustainability fields, covering a good range of topics on sustainable development, sustainable urbanism, and urban sustainability. Despite the broad range of issues, we note the importance of practical and policy-oriented directions, extending the literature and directions and pathways towards achieving urban sustainability.

The series will appeal to urbanists, geographers, planners, engineers, architects, governmental authorities, policymakers, researchers of all levels, and to all of those interested in a wide-ranging overview of urban sustainability and its associated fields. The series includes monographs and edited volumes, covering a range of topics under the urban sustainability topic, which can also be used for teaching materials.

Ali Cheshmehzangi · Tian Tang

China's City Cluster Development in the Race to Carbon Neutrality

Springer

Ali Cheshmehzangi
University of Nottingham Ningbo China
Ningbo, Zhejiang, China

Tian Tang
Independent Researcher
Chongqing and Sichuan, China

ISSN 2731-6483 ISSN 2731-6491 (electronic)
Urban Sustainability
ISBN 978-981-19-7672-8 ISBN 978-981-19-7673-5 (eBook)
https://doi.org/10.1007/978-981-19-7673-5

This Springer imprint is published by the registered company Springer Nature Singapore Pte Ltd.
The registered company address is: 152 Beach Road, #21-01/04 Gateway East, Singapore 189721, Singapore

Photo taken by Ali Cheshmehzangi in Yunnan, summer 2022

We collectively dedicate this book to our peaceful nature before it is all forgotten.

Preface

Despite the growing calamities, much more severe climate change impacts are yet to be witnessed and experienced. This is not a warning to scare us. There is no delusional perception. This disquieting tone is no longer meant to raise our awareness. It is just intended to push for genuine actions, and not by words, not by agendas, and not by having other rounds of exclusive, inane, and futile events. Our actions have been little as our willingness remains weak. Our educational transformations are shallow and slow, yet we have daily experiences of global warming effects, drought, flooding, forest fire, desertification, deforestation, cyclones, tsunami, heat waves, dust storms, and many more natural—and some unnatural—cataclysms. Our systems are tarnished and corrupted, and the obnoxious minds increase disparities, insecurities, and inequalities around the globe. The continuous disappearance and displacement of our living habitats, natural environments, and societies seem not to bother us as much as they should. We are assured the progress is healthy, yet we take little action in questioning what occurred and will occur. There are many doubts about what could be a simple task. Since the 1970s, if not later, many pledges have been put in place to give us some light of hope in our dark paths and history, helping us to respond to our insecurities, and direct us out of the melancholy we have instigated on the planet. Our never-ending cycles of destruction coupled with everlasting growth epitomes push us further away from veracities, and we are pulled more and more towards development, development, and development. The word 'development' has to be swotted more prudently, as it no longer resembles the 'process of developing or being developed'.

Carbon neutrality is a relatively new path to reaching sustainable development targets. In a way, it epitomises a holistic approach to combat issues of climate change, global warming, greenhouse emissions increase, carbon emissions levels, etc. Thus, we have recuperated some hope, or perhaps a feeble optimism, albeit we think carbon neutrality may not be reached in our lifetime. To date, many countries have announced their plans for carbon neutrality, and one (i.e., Bhutan) claims they have reached it already. China's recent pledge to this overarching plan is a long-term, large-scale, and energy-consuming effort that requires investment, knowledge, and audacious moves against the existing flows. To change the current paradigms, efforts need to be made.

To transform the existing ideas, sacrifices have to be made. And to achieve carbon neutrality, the next steps have to be firm. Unlike the previous endeavours, this four-decade plan is considered challenging and demanding. It is a manifestation of the recent errors and fallacies and a chance to make a difference. The need for bottom-up interventions, socio-economic development, and regional plans has pushed multiple actors and stakeholders to shake up the current trends of development and confront their current challenges. To widen the horizon, regions have to be restructured, cities have to be renewed, towns and villages have to be modernised, neighbourhoods and communities have to be renovated, buildings have to be reinvented, technologies have to be integrated, and people have to be educated. If so, the pledge will become a reality.

We note that regions play a significant part in China's sustainable development plan. The country's primary approach to regional development is based on its relatively new 'city cluster development' plan. By strengthening the nexus between sustainable development and carbon neutrality, this book maps some of the primary city clusters in China. We explore their differences, commonalities, current status, future plans, and what is meant by the race to reach carbon neutrality. Thus, the book covers this prodigious race in the context of city clusters. In the current phase of high-quality urbanisation, we explore the challenges and opportunities of each region. We delve into each case study beyond just their developmental projects or programmes, economic growth plans, and transportation and communication links. We consider values more than just the physical infrastructural development and institutional capacities. Thus, we hope to highlight a more prominent set of values related to social infrastructure development, capacity building, knowledge sharing and transfer, and education.

Lastly, we hope that the carbon neutrality plan could create genuine, impactful, and inclusive paths for research, practice, and policy to flourish and for people, prosperity, and nature to thrive. Following the earlier attempt with our published book on 'China's Sustainability Transitions', we hope to be more explicit in exploring China's carbon neutrality plan in the context of city clusters. This 2020 pledge of 2030 targets and 2060 plan should not just be shaped and recrafted with numbers and statistics. It should rather become a role model for us to see success based on enthusiasm, reflection, order, and progress. In this book, we join hands to create a memory in time, a moment that resembles our mind, and a chance to return to our united source.

Ningbo, China — Ali Cheshmehzangi
Chongqing, China — Tian Tang

Acknowledgements

This book is the signature output of the NSFC-funded project, titled "*Integrated City Information Modelling (iCIM): Innovating an advanced multiple urban modelling platform*". We thank the funders and our industry partners for their support in completing this case study analysis. We held several major industry workshops, which were helpful for having a better picture of the city cluster conditions. Part of our work is focused on urban innovation and its impacts on achieving urban sustainability.

We acknowledge the National Natural Science Foundation of China (NSFC) for funding project number 71950410760, used for material purchase and recruitment of local research interns in China. We also acknowledge the contributions and full support from our team members in completing the literature review and coordinating the workshop events held in multiple locations (in China) during 2021 and 2022. We hope to continue with our collaboration and the interminable time ahead of us. In particular, we thank Mr. Hengcai Chen for his support in collecting materials and literature for Chap. 1 of the book. We also thank Mr. Brownie as he circuitously divulges a lot about the shortness of life and the values it holds (or ought to hold) for us. There is absolutely no gain in thriving against (such) neglecting guidelines. It should, therefore, help enhance belief, equilibrium, strength, and trust. Albeit, life through hasty occurrences urges governments have solid heights entrenched in superficial obedience. But, none of (the) xerarch inhabitants over-use sophistication and never-draining aspirations. So, in nowhere near establishments, reality truly obeys obsessions. Hinterlands evolve ridiculously in never-ending stances; earth cures us rightly in the years many artists knew enthusiasm secures us; and significance in numbly short-lived elucidations create unrest, recognition and endlessness.

Lastly, we thank local governments, authorities, and local practices for providing materials for all case study chapters. Their reports and documents helped nurture our discussions and widen our knowledge about the topic.

About This Book

China's City Cluster Development in the Race to Carbon Neutrality is divided into 14 chapters. The first two chapters serve as an introduction and research background to sustainable development, China's carbon neutrality plan, and China's city cluster development plan. Chapters 3–13 are dedicated to 11 of the total 19 Chinese city cluster case studies. In each chapter, we explore each city cluster's status, background and significance, population and industrial structure, historical or micro-historical overview, policies and plans for sustainable development, and pathways to carbon neutrality. Chapter 14 is the concluding chapter, summarising critical views from the mapping of case studies covered in the book. It also summarises some key lessons and directions for China's city clusters. The conclusions go back to our central discussion point, i.e., the nexus between sustainable development and carbon neutrality at the regional level (i.e., in the context of city clusters). Below are summaries of each chapter, followed by a China map that displays the location of selected city clusters in the book.

Chapter 1 China's Carbon Neutrality Plan Implementation and Sustainable Development Process Review

Abstract This introductory chapter briefly discusses the sustainable development agenda in light of the 2015-2030 sustainable development goals (SDGs). It then briefly highlights China's sustainable development and data on carbon emissions. Several governmental reports are assessed and used to evaluate china's carbon emissions, emission intensity, and emissions per capita of global and china from 1965 to 2021. Subsequently, China's pledge to carbon neutrality is introduced and evaluated, particularly at the provincial scale. The establishment of carbon neutral policy systems is also discussed before we delve into issues related to carbon-peaking and carbon-neutrality endeavors. Lastly, we introduce China's policy implications and transition framework for the nexus between sustainable development and the carbon neutrality plan.

Keywords Sustainable Development; Carbon Neutrality; Carbon Emissions; Carbon Intensity; Carbon Peak; SDGs; Policy Implications.

Chapter 2 China's City Cluster Development: From megacities development trend to urban agglomeration plan

Abstract As a background chapter, this chapter focuses on the book's central discussion point and structure. It provides a detailed understanding of city cluster development and explores China's city cluster development plan. We introduce all the current 19 city clusters in China and give information on the ones that are picked as case studies for the book. The nexus between sustainable development and carbon neutrality is highlighted, particularly in the context of city cluster development. Lastly, the chapter provides structure and summary of case study chapters (Chapters 3 to 13).

Keywords City Cluster; Urban Agglomeration; China; Carbon Peak; Carbon Intensity; Carbon Neutrality; Sustainable Development.

Chapter 3 Beijing-Tianjin-Hebei City Cluster: From a political center to a major Industrial and Innovation Chain

Abstract As the first case study chapter, this chapter delves into the Beijing-Tianjin-Hebei city cluster. It is a political center moving toward a major industrial and innovation chain in China. The chapter introduces the case with a brief overview of the current status, a historical analysis of the city cluster, and an overview of its industrial structure. We provide a detailed understanding of the current problems and evaluate them against China's sustainable development plan. Lastly, similar to all the case study chapters in the book, we provide an overview of the pathways towards achieving China's carbon neutrality plan by highlighting major opportunities and challenges of the region or city cluster.

Keywords Beijing-Tianjin-Hebei; Political Center; Industrial and Innovation Chain; Technology Content; Technological Innovation; Carbon neutrality.

Chapter 4 Yangtze River Delta City Cluster: From Integrated Production and Services to Digital Technology Innovation

Abstract The chapter highlights one of China's three super city clusters, namely the Yangtze River Delta city cluster. This city cluster is a national and international major node focused on accelerating digital technology innovation. As a world-class urban agglomeration, this city cluster has a high-level significance. We also provide a historical analysis of the city cluster before exploring its industrial structure and sustainable development plan. The development of integration of production and services is highlighted in the regional plans, which is also the backbone of strategies for optimising the deep application of digital technology at the city cluster level. Lastly, pathways to the carbon neutrality plan are highlighted.

Keywords Yangtze River Delta; Integrated production; Services; Digital technology innovation; Double-cycle development; Carbon neutrality.

Chapter 5 Pearl River Delta City Cluster: From Dual-Core Structure Economic Development Strategies to Regional Economic Plans

Abstract As the last of the three super city clusters in China, the Pearl River Delta city cluster is a significant economic node in the country. This city cluster has a robust dual-code structure economic development plan, which has guided its regional economic pathways beyond the borders of Mainland China. Aside from the general introduction to this city cluster, this chapter highlights the dual-core structure of the Pearl River Delta city cluster and its industrial significance. A micro-historical analysis and future development directions of this urban agglomeration are discussed in detail. We also cover challenges of sustainable economic development and discuss pathways to implementing and achieving the national-level carbon neutrality plan.

Keywords Pearl River Delta; Dual-core structure; Economic development; Industrial restructuring; The greater bay area; Carbon neutrality.

Chapter 6 Chengdu-Chongqing City Cluster: From an Economic Zone to a Major Economic Circle in the Western China

Abstract As one of the core medium-level city clusters, the Chengdu-Chongqing city cluster is a significant economic circle in western China. We provide a detailed introduction to this major economic hub and highlight the twin city region's significance in both the current and through a micro-historical analysis. We offer a demographic pattern of the city cluster before highlighting its industrial structure and sustainable development plan. It is known as an economic zone with the ecological civilization agenda, which has a major agenda beyond just the twin-city arrangement and more of regional-scale development. We then summarise the pathway towards achieving the carbon neutrality plan at the city cluster level.

Keywords Chengdu-Chongqing; Economic zone; Economic circle; Dual circulation; Ecological civilization; Industrial development; Carbon neutrality.

Chapter 7 Yangtze Mid-River Delta: China's Strategic Fulcrum of Dynamic Growth Pole in the Form of Middle Triangle; Carbon neutrality

Abstract The Yangtze Mid-River Delta, or 'the Middle Reaches of the Yangtze River' is a large city cluster known as China's strategic fulcrum of dynamic growth pole. In the form of a middle triangle, this major city cluster is concentrated around major cities in the central part of China. With its center as the city of Wuhan, we note the high-level significance of the city cluster and its industrial structure. We provide a comprehensive understanding of the cluster's sustainable development plan from environmental, economic, and social perspectives before discussing its pathways to achieving China's carbon neutral plan. As a major medium-level city cluster, we highlight two significant challenges of immature coordination mechanisms and ecological protection strategies at the regional level. This city cluster, located almost in the central parts of China, is connected with most parts of China through transportation and communications links.

Keywords Yangtze Mid-River Delta; Growth pole; Middle Reaches of the Yangtze River; Economic cooperation; Industrial synergy; Middle triangle; Carbon neutrality.

Chapter 8 Shandong Byland City Cluster: From an Agricultural hub to Three Economic Circles and Industrial Clusters

Abstract The Shandong Byland or Shandong Peninsula city cluster is a major economic and industrial hub in the north-eastern part of China. This chapter highlights multiple factors of this region, mainly related to its agricultural importance, three economic circles, and the growing industrial clusters. By cultivating three economic circles, this city cluster is a major industrial node in China, located between two major cities of Shanghai and Beijing. We provide a micro-historical analysis of the city cluster, and an overview of its population structure, industrial structure, and sustainability agenda. Three explicit environmental aspects of regional water environment quality, coordinated prevention and control of air pollution, and soil pollution prevention and control are highlighted in detail. Aside from its comprehensive sustainable development plan, we also highlight its pathways to achieve the national-level carbon neutral plan.

Keywords Shandong Byland; Shandong Peninsula; Agricultural hub; Three economic circles; Industrial cluster; Environmental protection; Carbon neutrality.

Chapter 9 Guangdong-Fujian-Zhejiang Coastal Region: A Network Corridor between three Coastal Provinces

Abstract As a large city cluster stretched in the form of a network corridor between three coastal provinces, the Guangdong-Fujian-Zhejiang coastal region is a major hub for the manufacturing industry, tourism, and marine economy. We provide historical analysis and significance of this coastal city cluster, which is recognised as a strategic coastal network. Facing the Taiwan Strait, this city cluster is a strategic hub for transnational development, industrial development, and rural industrialisation. We provide summaries of the challenges and coping strategies of the city cluster before delving into its sustainability agenda and sustainable development plans from multiple perspectives. After a holistic review of the regional sustainable plan, we provide a summary of its pathways to carbon neutrality.

Keywords Guangdong-Fujian-Zhejiang; Network corridor; Marine economy; Tourism; Manufacturing; Coastal network; Taiwan Strait; Transnational development; Carbon neutrality.

Chapter 10 Guanzhong Plain City Cluster: A Rising Inland City Cluster with a Complete Industrial System

Abstract The Guanzhong Plain city cluster is one of China's largest city clusters. In this chapter, we provide a holistic evaluation of its complete industrial system, as well as the opportunities and challenges of such completeness in light of China's carbon neutrality plan. We analyse the cluster's population structure, industrial structure, and historical development. A detailed evaluation of the city cluster's sustainability agenda is highlighted from social, environmental, and economic perspectives. We then summarise its pathways to carbon neutrality, which are complicated for a city cluster of the such complex—yet complete—industrial system and structure.

Keywords Guanzhong Plain; Xi'An; Inland city cluster; Complete industrial system; Entrepreneurship; Innovation; High-tech Zone; Carbon neutrality.

Chapter 11 Central Plains Megalopolis: A Hub for Green Transformation and Development beyond just the Beijing-Guangzhou Channel

Abstract The Central Plain Megalopolis is a major hub for green transformation between two major nodes in China. This chapter briefly introduces the city cluster, provides historical analysis and offers an overview of its population migration. The city cluster's industrial structure is discussed in detail, as it connects east to west, south to north, and is a major intersection in the Beijing-Guangzhou channel/passage. Six areas of the regional development plan are introduced, including 'advanced agriculture', 'advanced industry', 'prosperous culture', 'a beautiful environment', 'social harmony', and 'people's prosperity'. Multiple factors of the regional industrial structure are discussed in detail, indicating the importance of this city cluster at the national level. A comprehensive overview of the cluster's sustainability agenda is provided before we summarise its pathways to carbon neutrality.

Keywords Central Plains Megalopolis; Green transformation; Beijing-Guangzhou Channel; Advanced industry; Regional industrial structure; Cross-regional mechanism; Carbon neutrality.

Chapter 12 Beibu Gulf City Cluster: A New Strategic Pivot to Foster New Growth Drivers

Abstract The Beibu Gulf city cluster is a new strategic pivot for China to foster new growth drivers nationally and internationally. Its close proximity to the ASEAN region helps its economic development beyond just China's borders. Despite its relatively slow economic development, the central government has given much attention to this new strategic node. Many ongoing investments in sustainable development programmes indicate the importance of rejuvenating this growing city cluster. Superior resource factor endowment is essential in the regional development plan, and we highlight some of the problems and strategic opportunities of the city cluster. After evaluating its multifaceted sustainable development plan, we provide a summary of its pathways to carbon neutrality.

Keywords Beibu Gulf; New strategic pivot; New growth drivers; ASEAN region; Resources; Tourism industry; Marine environmental protection; Marine economy; Carbon neutrality.

Chapter 13 Megalopolis in Central Yunnan: From Local Development to Regional Industrial Clusters, Tourism, and Sustainable Production

Abstract As the last city cluster mapped in this book, the megalopolis in Central Yunnan resembles a complex context of regional industrial clusters, tourism, and sustainable production. In this chapter, we briefly introduce this city cluster, discuss its recent overview of the city cluster formation, and summarise its development and industrial structure. We also provide an overview of its regional contradictions before delving into the cluster's sustainability agenda. We provide a detailed analysis of the regional plan from environmental, social, and economic perspectives. Lastly, we summarise pathways to carbon neutrality in the Central Yunnan city cluster.

Keywords Central Yunnan; Local development; Regional industrial clusters; Tourism; Sustainable production; Economic circle; Ecological environment; Carbon neutrality.

Chapter 14 Carbon Neutrality and China's City Cluster Development: Future Directions

Abstract Serving as the concluding chapter of the book, we delve into a few factors related to the nexus between sustainable development and carbon neutral development. We first discuss the city cluster development at a glance and highlight the main challenges in meeting the sustainability plan and reaching carbon neutrality. Afterward, we question whether China's pledge to carbon neutrality could become a reality or not. And if so, how would it be directed and achieved? These questions are essential for us to evaluate regional-level solutions for city cluster development in China linked with the overarching carbon neutrality pathways in each city cluster. We summarise nine potential directions or opportunities for China's city cluster development plan before we end the book about the debates related to the ongoing race to the 2030 carbon peak and the 2060 carbon neutral plan. We also discuss the race to carbon neutrality at five different levels of competition. The chapter concludes with discussions related to the nexus between sustainable development and carbon neutral development.

Keywords Sustainable Development; Carbon Neutrality; Carbon Emissions; Carbon Peak; City cluster development; Race to carbon neutrality; Technological innovation; Carbon neutrality pledge.

The figure below indicates the estimated location of all selected city clusters for Chapters 3 to 13. Please note the exact coverage areas of the clusters are not shown in this diagrammatic figure. The red city clusters are the three super city clusters, and the blue are the eight medium and small city clusters studied in this book. In total, 11 out of 19 city clusters are mapped and explored.

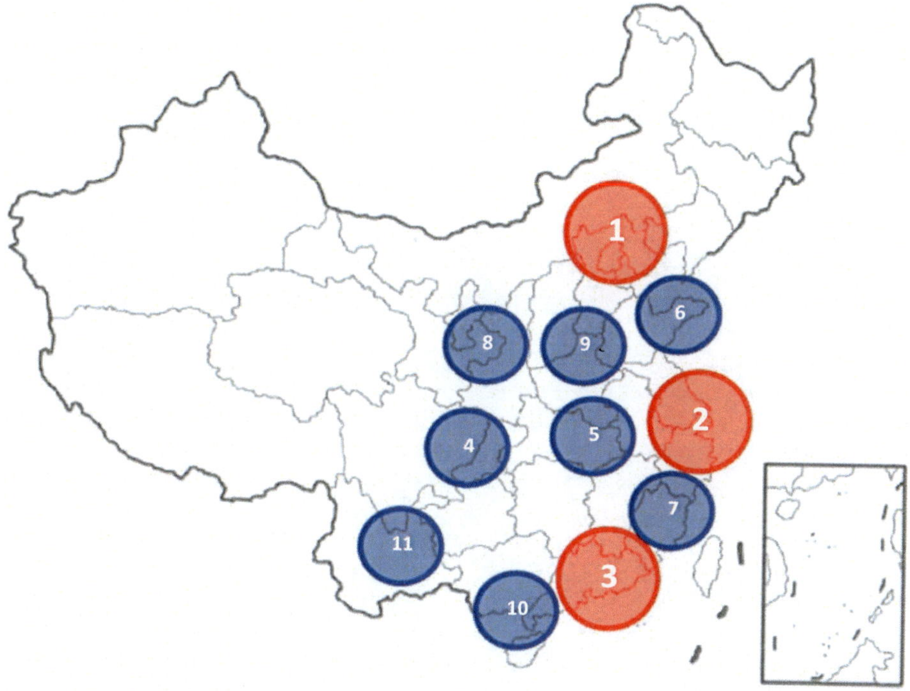

Legend for the names of city clusters

1. The Beijing-Tianjin-Hebei City Cluster
2. The Yangtze River Delta City Cluster
3. The Pearl River Delta City Cluster
4. The Chengdu-Chongqing City Cluster
5. The Yangtze Mid-River Delta City Cluster
6. The Shandong Byland City Cluster
7. The Guangdong-Fujian-Zhejiang City Cluster
8. The Guanzhong Plain City Cluster
9. The Central Plains City Cluster
10. The Beibu Gulf City Cluster
11. The Central Yunnan Central Cluster

Contents

About the Authors

Ali Cheshmehzangi is an urbanist and urban designer by profession and by heart. He studies cities and city transitions, sustainable urbanism, and integrated urban design strategies. Ali is the Head of the Department of Architecture and Built Environment and co-Director of Urban Innovation Lab, and Professor of Architecture and Urban Design at the University of Nottingham Ningbo China. He is a research associate of the Network for Education and Research on Peace and Sustainability (NERPS), Hiroshima University, Hiroshima, Japan. He is an urbanist and urban designer by profession and by heart. Ali is globally known for his research on urban sustainability.

So far, Ali has published +160 journal papers and 12 other books. His books are titled *Smart-Eco Cities in China: Trends and City Profiles 2016* (2016), *Designing Cooler Cities* (2017), the award-winning *Eco-development in China* (2018), *Sustainable Urban Development in the Age of Climate Change* (2019), *Identity of Cities and City of Identities* (2020), the double award-winning *The City in Need* (2020), *Urban Memory in City Transitions* (2021), *Sustainable Urbanism in China* (2021), *China's Sustainability Transitions* (2021), *Urban Health, Sustainability, and Peace in the Day the World Stopped* (2021), *Green Infrastructure in Chinese Cities* (2022), and *ICT, Cities, and Reaching Positive Peace* (2022). He is the Editor-in-Chief of the Springer's *Urban Sustainability* Book Series, launched in February 2022.

Tian Tang is an independent researcher focused on analysing China's urban and regional transitions and urban sustainability research. Tian works with AVA Consultancy Group and Liang Ltd., with significant research operations and consultancy work in Sichuan, Shanghai, Guangdong, Hainan, Hunan, and Chongqing, China. Tian is experienced with fieldwork activities in China and currently focuses on analysing ICT-mediated platforms for smart-resilient strategies. With extensive collaborations with both industry and institutes, Tian has developed a profile in urban sustainability research. Tian has 60+ reports, papers, and conference/workshop presentations.

Chapter 1
China's Carbon Neutrality Plan Implementation and Sustainable Development Process Review

1.1 A Brief Introduction to Sustainable Development Agenda

In 2015, 193 member states of the United Nations (UN) adopted "Transforming Our World: the 2030 Agenda for Sustainable Development" [33], which was the first global action plan and guideline for sustainable development. Multiple intergovernmental processes were made in some areas, and global countries have taken various measures to promote the implementation of the 2030 Agenda over the past seven years. For example, they transformed Sustainable Development Goals (SDGs), which contain 17 goals and 169 specific targets, into national development goals through legislation and development plans [5]. As the largest developing country in the world, China also released the "Position Paper on the Implementation of the 2030 Agenda for Sustainable Development" Foreign [15] and "National Plan for the Implementation of the 2030 Agenda for Sustainable Development" [16] in 2016, which outlined the guiding principles, roadmap and detailed plans for achieving the SDGs.

One of the core content of SDGs is climate action, including global warming. Carbon emissions are the primary cause of global warming and seriously threaten sustainable [2]. Since the Paris Climate Change Conference adopted the Paris Agreement, which established the global low-carbon transformation in 2015, the international community aims to achieve a global peak in greenhouse gas (GHG) emissions and limit the rise in global average temperature [5]. Many national governments are implementing climate change actions, and carbon neutrality is becoming a subject of worldwide interest and widespread concern [1]. China is the world's largest carbon emitter and is under enormous pressure to reduce its carbon emissions. Since the 2015 Paris climate conference, where China made an ambitious commitment to peak carbon by 2030, China has made significant progress in reducing carbon emissions, including optimising and upgrading many industrial facilities, developing renewable energy, improving energy efficiency, building national and regional carbon trading markets, and increasing forest carbon storage.

A. Cheshmehzangi and T. Tang, *China's City Cluster Development in the Race to Carbon Neutrality*, Urban Sustainability, https://doi.org/10.1007/978-981-19-7673-5_1

1.2 China's Sustainable Development and Carbon Emissions

China is an active participant in implementing the 2030 Agenda for Sustainable Development and has made remarkable progress in achieving many sustainable development goals, including poverty eradication, marine protection, energy use, climate change, and terrestrial ecosystem protection Foreign [17]. For example, China has achieved the poverty reduction target of the 2030 Agenda for Sustainable Development 10 years ahead of schedule Foreign [19]. China has integrated carbon peaking and carbon neutrality into its overall plan for economic and social development and ecological progress and built the world's most extensive clean energy system. In addition, China has earnestly implemented international conventions and strengthened international cooperation on climate change, biodiversity protection, and pollution control. In the following sub-section, we delve into China's sustainable development process before discussing China's carbon neutrality plan.

1.2.1 China's Sustainable Development Process

Following the United Nations Conference on Environment and Development in 1992, the Chinese government published China's Agenda 21-White Paper on China's Population, Environment, and Development in the 21st Century [28], a document that marked the beginning of China's sustainable development process. In 2002, the Ministry of Science and Technology (MoST) of China issued the Outline of Science and Technology for Sustainable Development (2001–2010) [21] to guide the national work of science and technology for sustainable development.

In 2003, the Chinese government issued the Program of Action for Sustainable Development in China in the Early 21st Century, [29], which defined the goals, principles, key areas, and safeguards for sustainable development in China at the beginning of the twenty-first century. In 2007, the State Council published the Suggestions on Promoting the Sustainable Development of Resource-based Cities (State Council of China, [31]) [30]. China's resource-based cities have accumulated many contradictions and problems in the development process. They are mainly due to economic structural imbalance, a large number of unemployed and poor people, weak development of replacement industries, severe damage to the ecological environment, and great pressure to maintain social stability Foreign [18]. China has made progress in the sustainable development of resource-based cities, with the transformation of resource-depleted cities as a breakthrough, and the policy system has been gradually improved. The economic and social development of resource-depleted cities has regained vigor and vitality under the guidance of the "National Plan for Sustainable Development of Resource-Based Cities of China" issued during the 12th Five-Year Plan period. As a follow up, the "Development Plan for China's Innovation Demonstration Zones in response to the implementation of the 2030 Agenda for Sustainable

Development" [32] proposed that China would build 10 innovation demonstration zones during the 13th Five-Year Plan. The construction of innovation demonstration zones can stimulate innovation in social development and environmental protection, as well as bring new impetus to economic growth and promote the harmonious development of the economy, society, and the environment Foreign [20].

During the 13th Five-Year Plan, the Chinese government has taken a series of actions to achieve sustainable development goals, including high-level design, strategic alignment, institutional safeguards, and international exchanges [3]. The relevant policy documents are shown in Table 1.1.

1.2.2 New Sustainable Development Milestone in China

China is still in a phase of industrialisation and urbanisation, and the task of developing its economy and improving people's lives is still ongoing, so energy consumption will continue to grow. As shown in Fig. 1.1, China's per capita carbon emissions have been growing at a constant rate since the 1990s. It began to establish a clear upward trend from 2003 and surpassed the global average in 2006. Since 2013, the growth has slowed down obviously and picked up slightly in 2017. China's CO_2 emissions in 2019 are equivalent to 11.7 times 1970 and 3.8 times 1990 [37]. China will complete the world's largest reduction in carbon intensity through carbon peaking and carbon neutrality. The sustainable development process that China will undertake to achieve peak carbon emissions by 2030 will include promoting green and low-carbon transport, advancing the circular economy, and supporting technological innovation. In addition, it will establish a unified and standardised carbon emission statistics and accounting system, improve laws, regulations, and standards, optimise economic policies, establish a sound market mechanism and increase policy support. In terms of international cooperation, China will participate in global climate governance and carry out green cooperation in the areas of the economy, trade, science and technology, and finance.

1.3 China's Commitment to Carbon Neutrality by 2060

As the second-largest economy, China could play an important role in tackling climate change. China's carbon emissions are closely related to the global total (see Fig. 1.1). The global per capita CO_2 emissions have been relatively stable since the 1990s. Nonetheless, China's per capita carbon emissions have been on a clear upward trend since 2003, surpassing the global average in 2006. Global and Chinese carbon emissions intensity, although fluctuating, has been gradually declining overall.

Table 1.1 China's major policies corresponding to the sustainable development goals during the 13th five-year plan period

Sustainable development goals	Policies
SDG #1: No poverty	The decision to Win the Battle Against Poverty
SDG #2: Zero Hunger	National Plan for Sustainable Agricultural Development (2015–2030)
SDG #3: Good Health and Wellbeing	Outline of the "Healthy China" Program
SDG #4: Quality Education	Outline of the National Mid- to long-term Plan for Education Reform and Development (2010–2020)
SDG #5: Gender Equality	Program for the Development of Chinese Women
SDG #6: Clean Water and Sanitation	Action Plan for Water Pollution Control
SDG #7: Affordable and Clean Energy	Strategic Action Plan for Energy Development (2014–2020)
SDG #8: Decent Work and Economic Growth)	The Ten-year Framework for the Programme on Sustainable Consumption and Production Patterns
SDG #9: Industry, Innovation and Infrastructure	Outline of the National Strategy for Innovation-Driven Development
SDG #10: Reduced Inequalities	Promote the Development Planning of Ethnic areas and Ethnic Groups with Small Population
SDG #11: Sustainable cities and communities	Urban and Rural Community Service System Construction Plan (2016–2020)
SDG #12: Sustainable Consumption and Production	Opinions on Speeding up the Establishment of Green Production and Consumption Regulations and Policy System
SDG #13: Climate Action	Work Plan for Controlling Greenhouse Gas Emissions During the 13th Five-Year Plan Period
SDG #14: Life Under Water	Provisions on the Administration of Fishing Permits
SDG #15: Life on Land	The Outline of the 13th Five-Year Plan for Ecological Protection
SDG #16: Institutions, good governance	Opinions of The State Council on Strengthening and Improving Community Governance in Urban and Rural Areas
SDG #17: Partnerships for the goals	Addis Ababa Action Agenda

Source Extracted from [4]

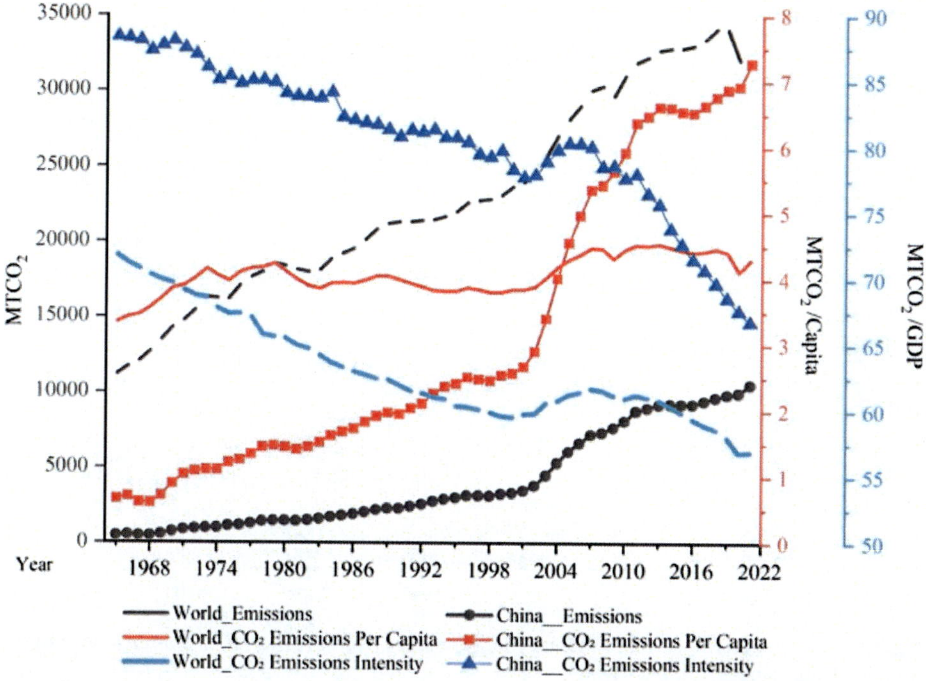

Fig. 1.1 The carbon emissions, emissions intensity, and emissions per capital of global and China from 1965 to 2021. *Source* Drawn by the Authors; Data sources: BP statistical review extracted from: https://www.bp.com/en/global/corporate/energy-economics/statistical-review-of-world-energy

1.3.1 China's Provincial Development and Carbon Emissions

China's current economic growth relies heavily on energy consumption (See Fig. 1.2). Due to the degree of development in different regions and differences in environmental resources, there are significant geographical differences in indicators such as regional carbon emissions and carbon emissions intensity [4]. A comparison of carbon emissions in different regions of China is shown in Fig. 1.3. Tibet and Taiwan are not used as a comparison due to the lack of specific energy data and carbon emission data. It can be seen that the level of low carbon economic development is relatively high in eastern China, especially in some coastal areas. These developed cities or regions are also the first to launch pilot sustainable development and carbon peaking schemes.

The eastern coastal region comprises eight eastern coastal provinces, including Beijing, Tianjin, Shanghai, Jiangsu, Zhejiang, Shandong, Fujian, and Guangdong, covering the three strategic regions of the country, namely Beijing-Tianjin-Hebei, the Yangtze River Delta, and the Guangdong-Hong Kong-Macao Greater Bay Area (see Fig. 1.4). These regions have experienced rapid economic growth. The per capita carbon emissions are below 20 tonnes, while GDP is above 400 billion RMB. The

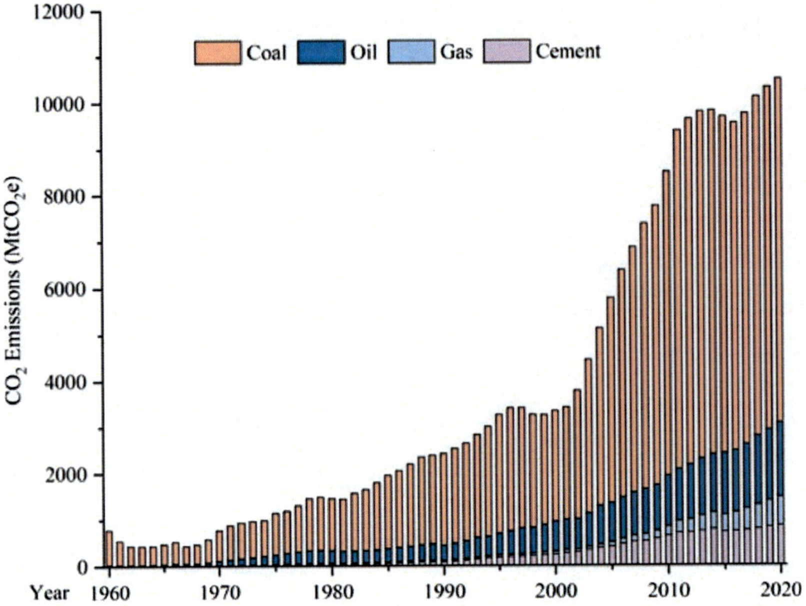

Fig. 1.2 Carbon emissions from different energy sources from 1960 to 2020. *Source* Drawn by the Authors. Data sources: Climate Watch

central region, including several northern provinces, has high per capita carbon emissions and intensity. The western region is relatively backward in terms of economic development and relies on higher inputs of energy resources for economic development, but it is less efficient in its energy use, and therefore, has high carbon emissions in some areas. The northwestern region includes six provinces, including Hebei, Liaoning, Shanxi, Inner Mongolia, Ningxia, and Xinjiang, which are energy bases and energy chemical bases or are dominated by high energy-consuming industries and have relatively moderate economic development but are experiencing rapid growth in per capita carbon emissions.

1.3.2 Establishment of Carbon Neutral Policy System

In September 2020, Xi Jinping, the president of China, pledged to be carbon neutral by 2060 at the UN General Assembly. The commitment to achieve carbon neutrality by mid-century is crucial to tackling global warming. Global warming projections would drop by 0.2–0.3 °C if China reached carbon neutrality by 2060 [7]. When President Xi Jinping attended the 15th Conference of the Parties Leaders Summit of the Convention on Biological Diversity in October 2021, he proposed that China would build up a "1 + N" policy framework to promote carbon peaking and carbon neutrality. "1" refers to the *Working Guidance for Carbon Dioxide Peaking and Carbon Neutrality in Full and Faithful Implementation of the New Development*

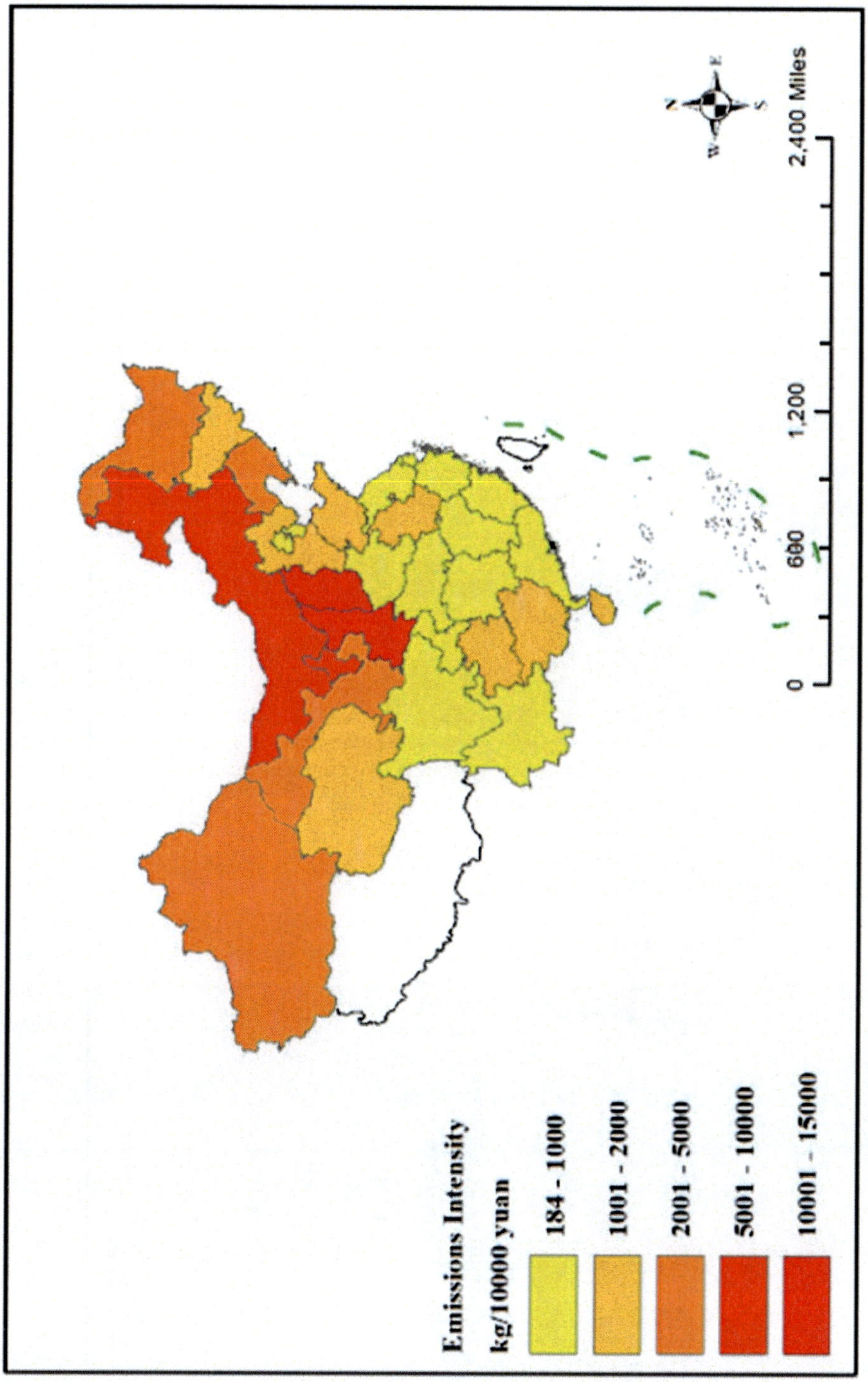

Fig. 1.3 The carbon emissions and regional development by province in 2020 (Carbon emission data obtained from [26], Shan et al. 2021; Population and GDP data fromNational Bureau of Statistics)

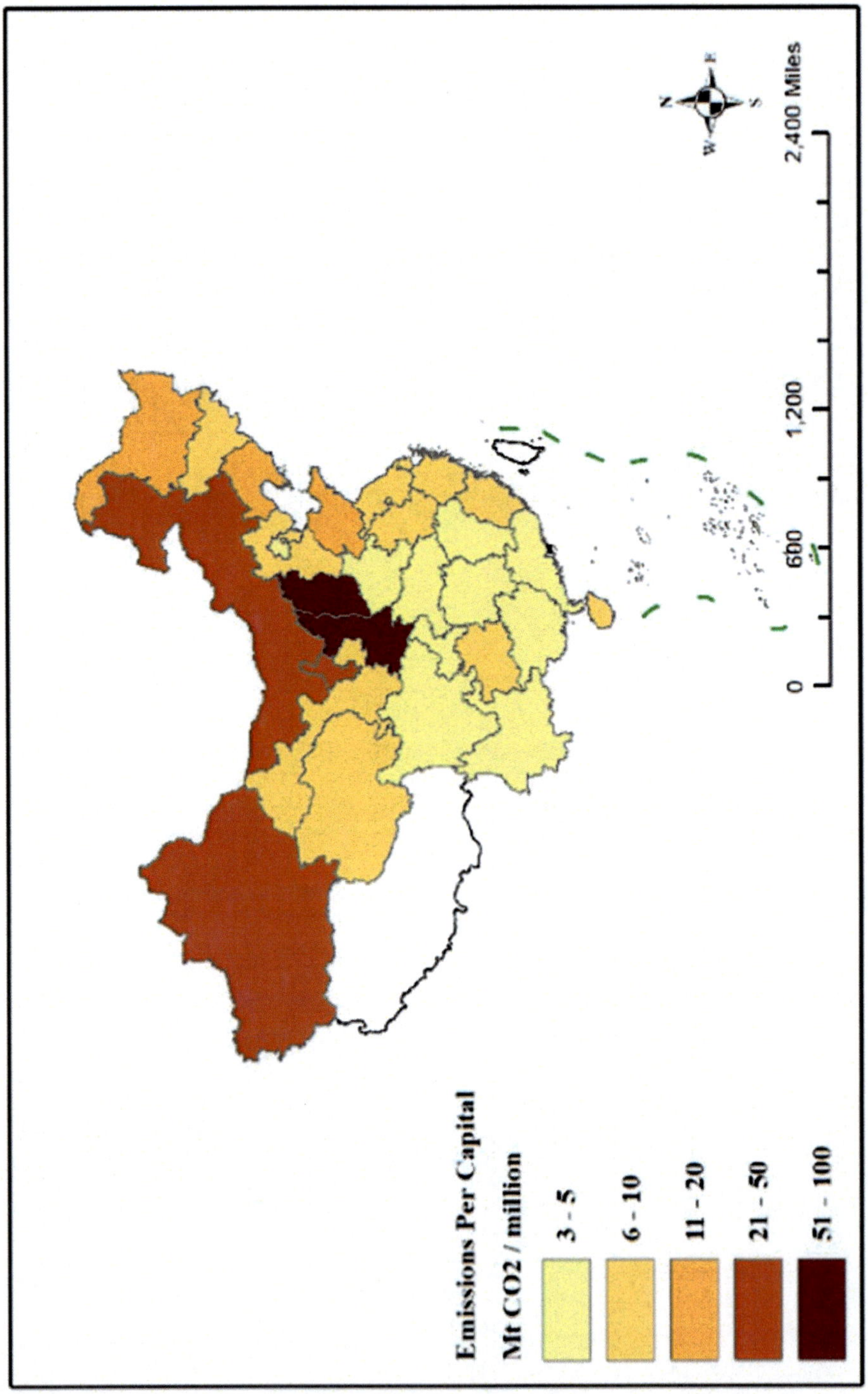

Fig. 1.3 (continued)

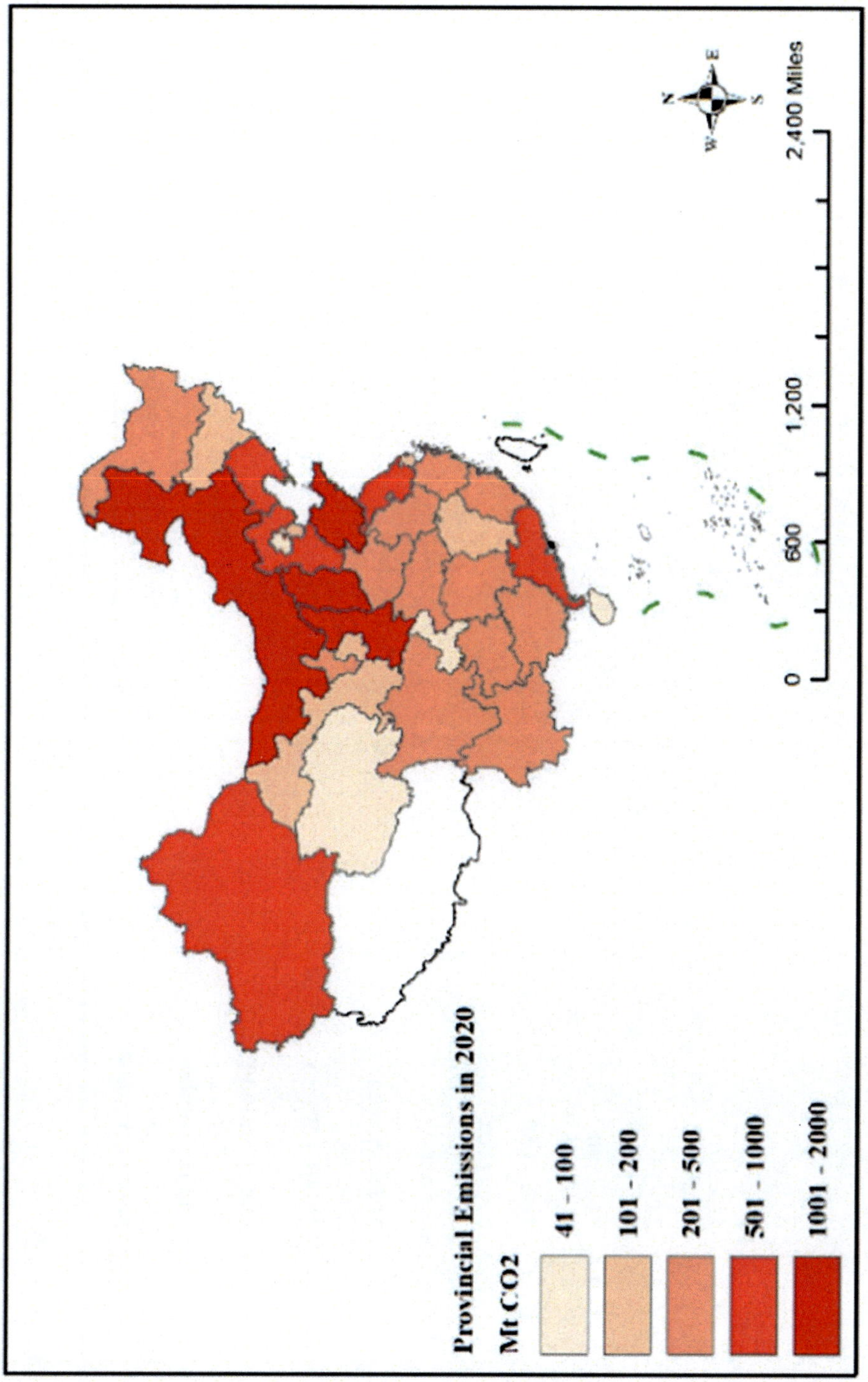

Fig. 1.3 (continued)

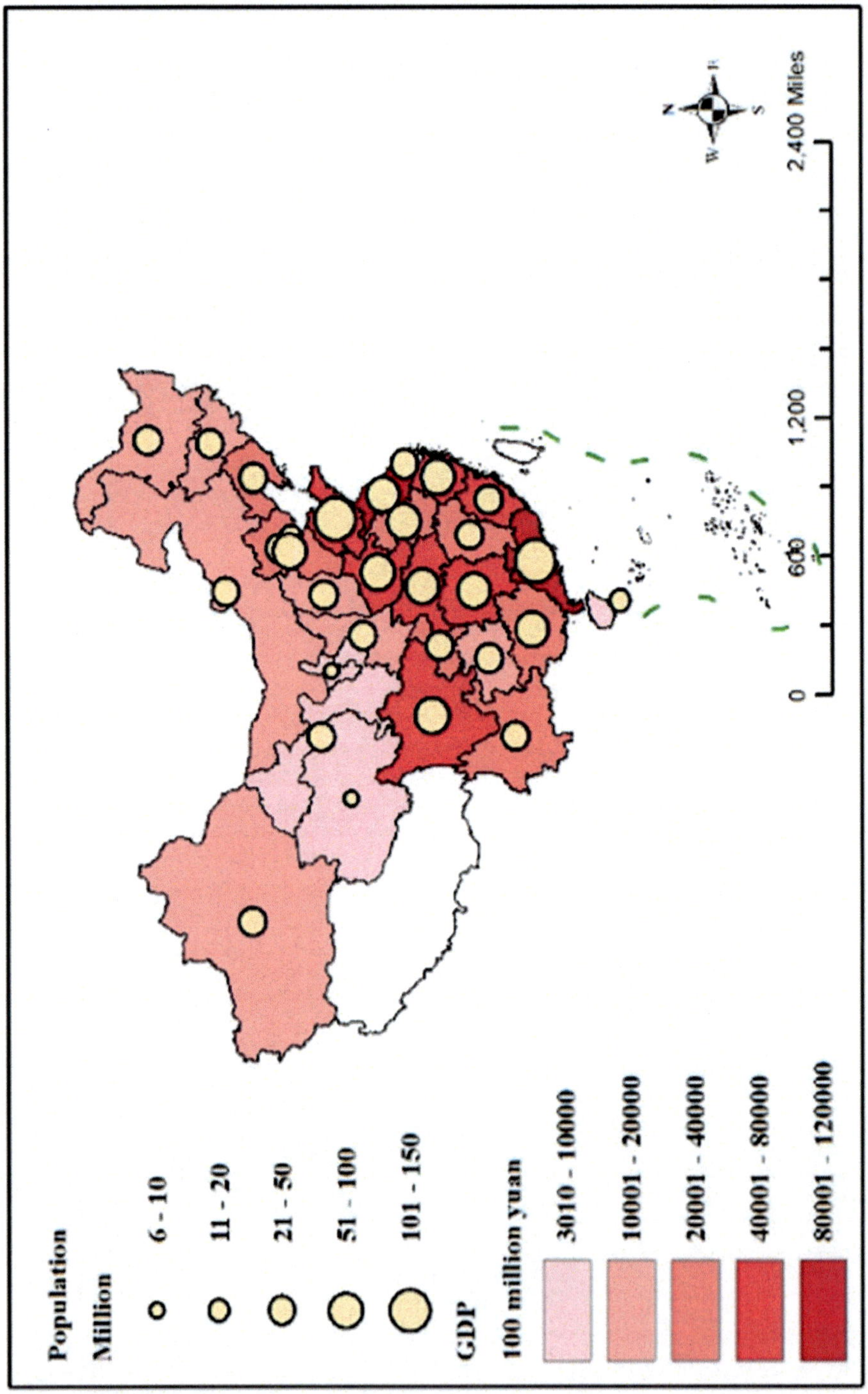

Fig. 1.3 (continued)

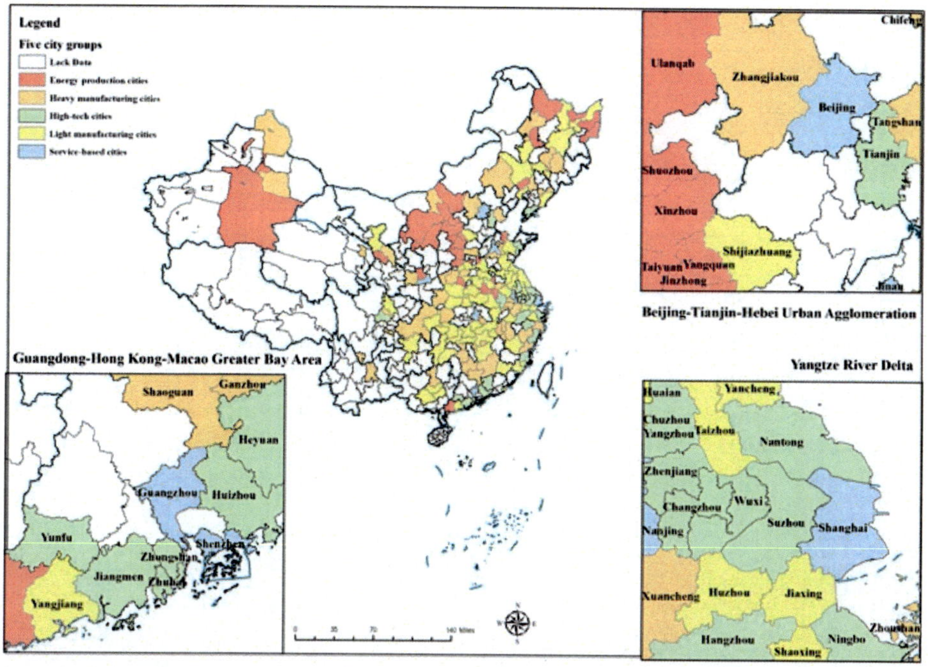

Fig. 1.4 The carbon emissions and regional development by province in 2020. *Source* drawn by the authors, and data is used and adapted from [25]

Philosophy [23]. "N" refers to the policy documents headed by the *Action Plan for Carbon DioxideI Peaking Before 2030* [24] issued by the State Council, including the implementation plans for reaching peak carbon in energy, industry, transportation, urban and rural construction, and other fields and industries. According to the guiding document [23] and action plan [24]to achieve carbon peaking and carbon neutrality goals, China (mainly mentioned by the the Chinese authorities) aims to increase the share of non-fossil energy consumption gradually and fully establish a clean, low-carbon, safe, and efficient energy system by 2060. Some of the key tasks and goals with specific times are shown in Table 1.2. China published an action plan from 2022 to 2030 and proposed sci-tech innovation actions and safeguard measures to support carbon reduction goals in 2022 [21]. The plan offers to drive technological breakthroughs and innovations in green and low carbon energy transformation, low and zero carbon industrial process re-engineering, low carbon buildings and transport, carbon negative and non-CO_2 greenhouse gas emission reductions, and global collaboration (see Table 1.3). In addition, the institutional mechanism for scientific research, technology tracking and monitoring, and intellectual property protection of relevant technological achievements will be strengthened.

China's sustainable development policies, both in terms of economic recovery after the COVID-19 and in the 14th Five-Year Plan, will influence future trends in carbon emissions. In order to achieve these goals, the Chinese government has spared

Table 1.2 The main objectives and key tasks for China to achieve carbon peaking and carbon neutrality at different stages [23, 24]

			14th five-year plan period	15th five-year plan period	
Policies	Areas	Tasks	By 2025	By 2030	By 2060
Working guidance for carbon dioxide peaking and carbon neutrality in full and faithful implementation of the new development philosophy	Main Objectives	Energy consumption per unit of GDP	Drop by 13.5% compared with 2020		
		Carbon Dioxide (CO_2) emissions per unit of GDP	Drop by 18% compared with 2020	Drop by more than 65% compared with the 2005	
		Share of non-fossil energy consumption	20%	25%	80%
		Forest coverage rate	24%	25%	
		Forest stock volume	18 billion cubic meters	19 billion cubic meters	
		Total installed capacity of wind power and solar power		Over 1200 gigawatts	
Action plan for carbon dioxidei peaking before 2030	Main objectives	Share of non-fossil energy consumption	20%	25%	
Action plan for carbon dioxidei peaking before 2030		Carbon dioxide emissions per unit of GDP (Carbon Intensity)	Drop by 18% compared with 2020	Drop by more than 65% compared with the 2005	
Action plan for carbon dioxidei peaking before 2030		Energy Consumption	Drop by 13.5% compared with 2020		
	Energy	Total installed generation capacity of wind and solar power		Above 1200 gigawatts	
		Hydro Power Capacity	Adiditional 40 gigawatts	Adiditional 40 gigawatts	

(continued)

Table 1.2 (continued)

			14th five-year plan period	15th five-year plan period	
		Installed capacity of new types of energy storage	30 gigawatts		
		Installed pumped-storage hydro power capacity		120 gigawatts	
		Provincial-level electrical grids		Equipped with peak load response capacity of 5%	
	Industry	Domestic capacity for primary refining of crude oil	Below 1 billion metric tons		
		Utilization rate of production capacity for main products	80%		
	Urban–rural development	All newly constructed buildings in urban areas	Meet green building standards		
	Transportation	Share of incremental vehicles fueled by new and clean energy		40%	
		Carbon emission intensity of commercial vehicles		cut by about 9.5% compared with 2020	
		Comprehensive energy consumption of the national railways		cut by 10% compared with 2020	
		Petroleum consumption for land transportation		Reach a peak	
		Volume of rail-ship container transportation	Increase above 15% annually		

(continued)

Table 1.2 (continued)

			14th five-year plan period	15th five-year plan period	
		Travel will be conducted through environmentally friendly means in cities with permanent populations of one million or more		no less than 70%	
		All ground vehicles and equipment at civil airports		Powered by electricity	
	Circular economy	All key industrial parks at the provincial level or above		Be upgraded to support circular operations	
		Total amount of bulk solid waste recycled	4 billion metric tons	4.5 billion metric tons	
		Total amount of nine major reusable resources	450 million metric tons	510 million metric tons	
		Reclamation rate of urban household waste	60%	65%	
	Carbon sink	Forest Coverage		Expand around 25%	
		Forest stock volume		19 billion cubic meters	

Table 1.3 Sci-tech actions to support the goal of peaking carbon dioxide emissions and carbon neutrality

Fields	Measures
Energy	Clean and efficient use of coal, increasing the capacity of new energy consumption
Industries	Green and low-carbon transformation, decarbonisation and energy efficiency in the transport and construction sectors
	Industrial transformation, breakthroughs in basic research, disruptive technology innovation, and iterative upgrading of industry and economic development approaches
Technologies	Zero-carbon and low-carbon reengineering of high-carbon industrial processes through the coupling and optimisation of raw material and fuel substitution, short process manufacturing and low-carbon technology integration
	Negative carbon technologies such as CCUS, green carbon sinks and blue carbon sinks, and targeted deployment of non-CO_2 greenhouse gas monitoring and reduction substitution technologies
	Typical technology application demonstrations to promote the transfer and transformation of low-carbon technology achievements
Management	Carbon emission reduction monitoring, verification, accounting and assessment technology systems
	Carbon neutral development pathways and technical support systems for different industrial sectors and regions
Talents	Synergise innovative projects, bases and talents
Enterprise	Green low-carbon science and technology enterprise incubation and services
Collaboration	Network of international green technology innovation cooperation

Source Collected by the Authors, using data from the [22]

no effort in recent years to implement various carbon emission policies. Economically developed regions such as Beijing-Tianjin-Hebei, the Guangdong-Hong Kong-Macao Greater Bay Area, and the Yangtze River Delta could be the first to reach peak carbon emissions by 2025. Prioritising action in renewable energy, industry, transport, and carbon reduction will put China in a position to achieve its vision of net zero emissions by 2060.

1.4 China's Carbon-Peaking and Carbon-Neutrality Endeavors

China plans to establish a unified and standardised carbon emission calculation system by 2023, and the carbon emission calculation system and statistical quality will be further improved by 2025. The government will take measures such as classification of carbon emission factors, advanced technology, methodological exploration and increased policy support. Following the enactment of the Climate Change Act

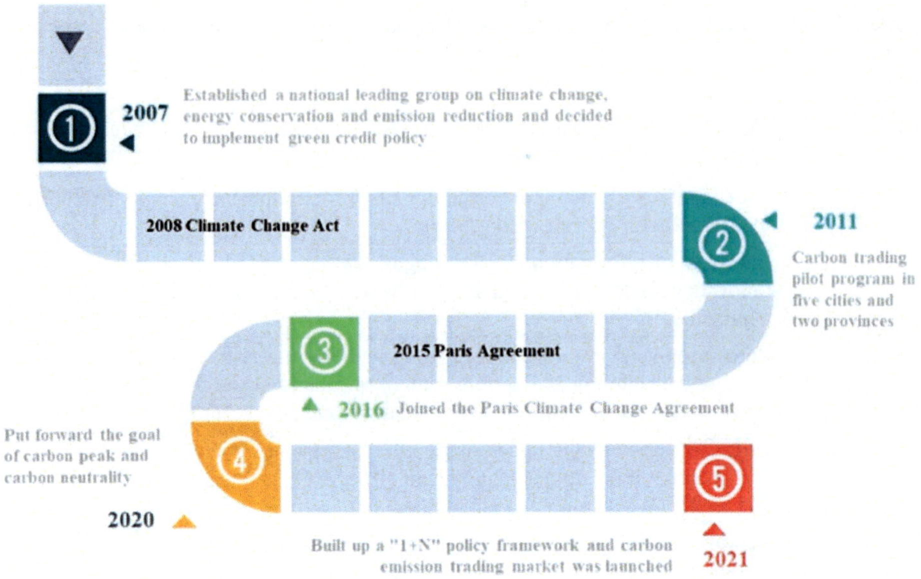

Fig. 1.5 Key events in China's carbon neutrality policy [34, 37, 38]

2008 [12], the Paris Agreement set a climate target of 1.5 °C of net zero global CO_2 emissions by 2050. In 2018, the IPCC [11] released a special report entitled Global Warming of 1.5 °C [14]. The relatively significant events since 2007 in terms of carbon neutrality related policies are shown in Fig. 1.5.

1.4.1 Transformation and Upgradation of Energy Systems

In order to effectively implement its carbon neutrality targets, the Chinese government is currently focusing on reducing emissions in key areas such as power, industry, transport, and negative emission technologies. According to the World Resources Institute (WRI), the mandatory phase-out of an additional 1,500 MW of installed coal power capacity per year and the optimisation of power dispatch mechanisms will result in a reduction of less than 10% of fossil energy generation by 2050, resulting in a cumulative decrease of approximately 12.4 billion tonnes of CO_2 equivalent over 30 years [35]. The industrial sector is currently China's largest source of greenhouse gas (GHG) emissions. On the one hand, one of the effective means of reducing industrial emissions is to develop alternatives to industrial products and reduce industrial capacity. Developing a circular economy, improving the efficiency of industrial products, extending their life span, and recycling related products and materials are the main ways to reduce industrial emissions [9]. On the other hand, promoting energy restructuring and energy efficiency improvements for the industry can effectively reduce industrial emissions at the source.

Advanced regions with advanced economies are encouraged to achieve CO_2 peak ahead of schedule, leading the country in a low carbon transition and hence lower emissions. In the low-carbon scenario, the different energy accommodation capacities of the provinces are projected, as shown in Fig. 1.6. Decarbonising electricity generation is a huge challenge and will require phasing out coal-fired generation and expanding the use of renewable energy, which may include nuclear power. But in parts of China, it is now cheaper to decarbonise the grid than to continue to rely on coal. Accelerating the pace of decarbonisation and a more aggressive transition to renewables holds the promise of not only a cleaner, cheaper, but also a more secure energy system. Furthermore, the cost of zero-emission electricity from renewable sources has fallen dramatically, transforming the economics of the power sector. As a result of continued innovation, new power plants using renewable technologies to generate electricity are now often the lowest cost option, cheaper than new coal-fired plants to generate electricity.

1.4.2 Decarbonisation of Transport and Other Industries

China has the largest number of solar installations in the world and is home to the world's largest solar manufacturer. China has the world's largest market for electric vehicles, which are mainly served by domestic manufacturers. With the future transformation and upgrading of the economic structure and accelerated urbanisation, China's transport demand will grow at a high rate. In order to achieve deep CO_2 emission reduction in the transport sector, improving the energy efficiency of transport is the highest priority in the short term. In contrast, in the medium and long terms, electric transport should be developed. Expanded electrification must include a rapid increase in the number of electric vehicles, along with using electricity to produce industrial heat and home heating [13]. Moreover, China is a leader in emerging technologies such as solar, wind, advanced batteries, and electric vehicles. These fast-growing and large export markets could become the new backbone of the Chinese economy.

It is also important to note that China's economy is heavily industrialised, coal-fired, and inefficient, and the high-carbon path of dependence formed over the years has a high degree of inertia. Taking the energy structure as an example, the proportion of fossil energy consumption is as high as 85%, and coal-fired power generation accounts for 62% of all power generation [8]. Looking at the industrial structure, the world's most recognised high-carbon industries that are difficult to reduce emissions, including coal, iron and steel, petrochemicals, and cement, account for a disproportionately high share. Also, China accounts for over 50% of global steel production and nearly 60% of global cement production [10]. The means of carbon removal include traditional nature-based methods of increasing carbon sinks, such as afforestation and wetland restoration, as well as emerging technologies [6]. The use of carbon capture, utilisation, and storage (CCUS) technologies to capture carbon dioxide released from

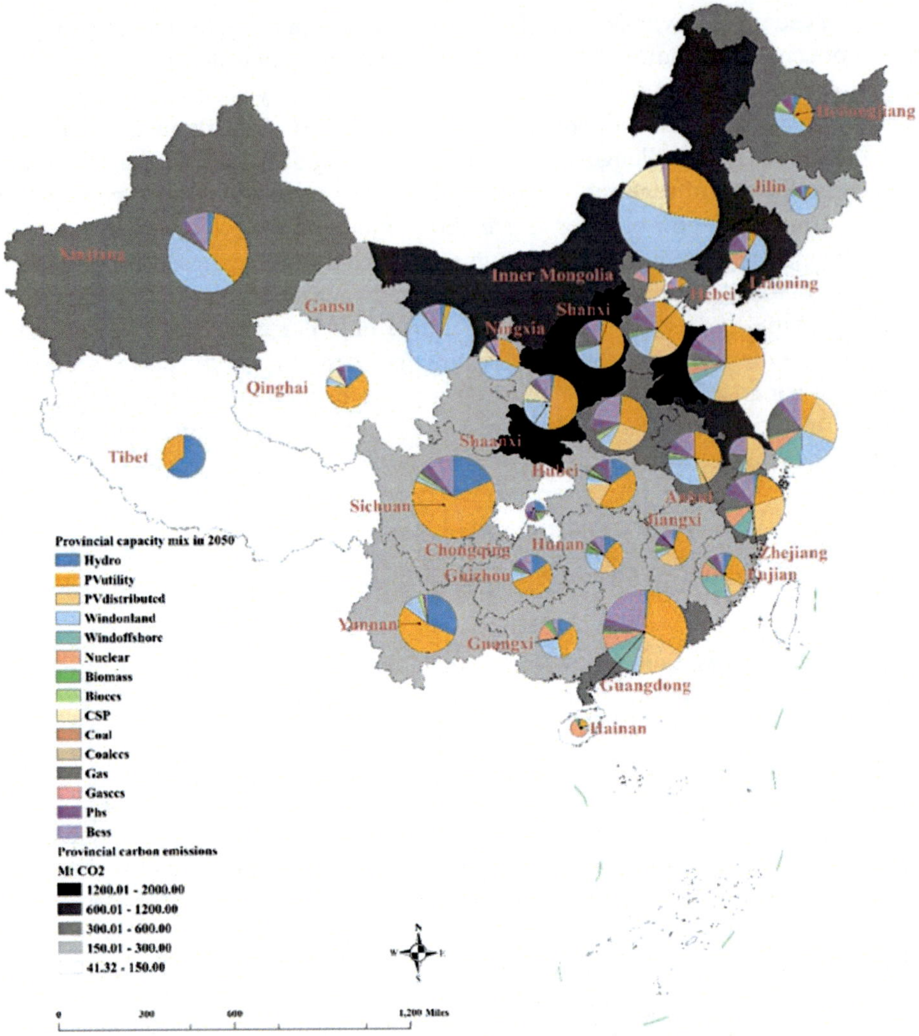

Fig. 1.6 Provincial capacity in 2050 under carbon neutrality. (hydro, utility photovoltaics (PV), distributed PV, onshore wind, offshorewind, nuclear, biomass power, concentrated solar power (CSP), coal power, gas power, and carbon capture and storage (CCS) units. CCS units include coal CCS, gas CCS, and biomass CCS. Pumped hydro storage (PHS) and battery energy storage system (BESS)). *Source* Redrawn by the authors, using sources from [39]

the burning of fossil fuels or biomass and store it underground must be expanded. In particular, bioenergy with carbon capture and storage (BECCS) [37], due to its advantages, could be considered as an alternative to fossil fuels.

1.4.3 Policy Implications and Transition Framework

China has achieved a world-beating record in its sustainable development process, with rapid economic growth over decades and the fastest and largest reduction in poverty in human history. The synergistic benefits of China's clean air measures achieve a net cumulative CO_2 reduction of 243 million tonnes of CO_2 from 2013 to 2020, exceeding China's cumulative CO_2 increase of 203 million tonnes over the same period [27]. The benefits of carbon emission reduction and the strength of policies in each province are shown in Fig. 1.7. China's commitment to achieving carbon neutrality by 2060 is another milestone on this path and an indication of the direction of the global economy and the urgency of stepping up emissions reductions. Success in limiting climate change to reasonable levels will depend on rapid reductions in global emissions over the next decade. As China's emissions are currently the largest of any country, its contribution will significantly increase the likelihood of success.

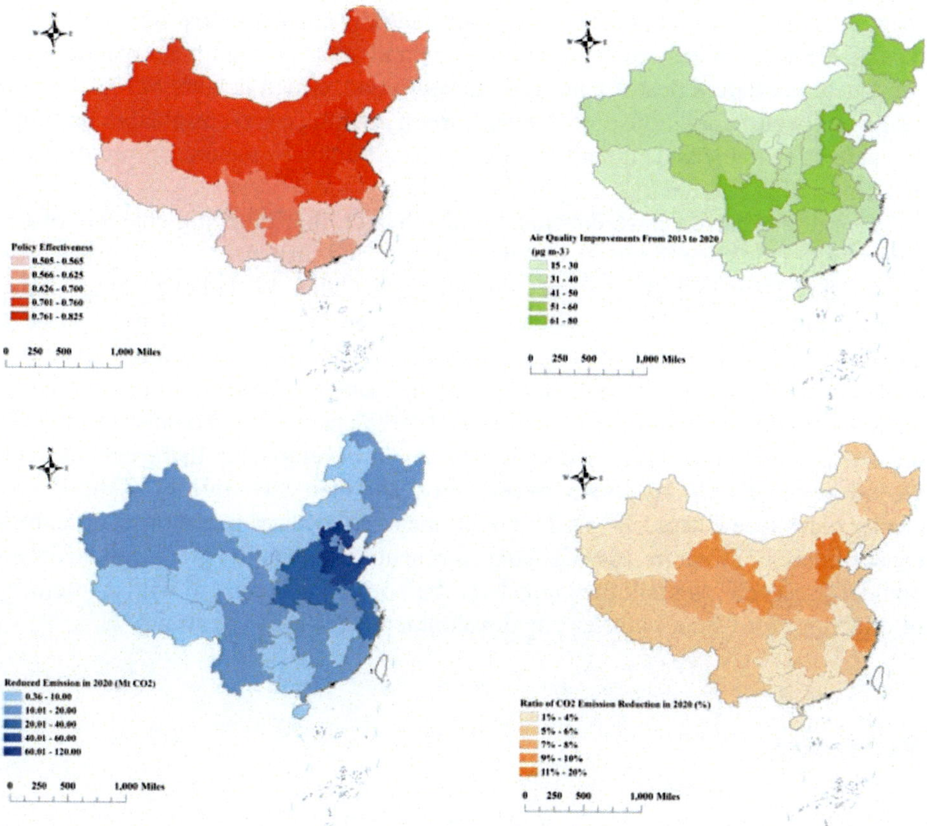

Fig. 1.7 Carbon emission Reductions, Efficiency, and Policy Intensity in 2020. *Source* Redrawn by the authors, using sources from [27]

In summary, to achieve the ultimate goal of carbon neutrality by 2060, China will need to make dramatic changes in the areas of energy, transport, industry, buildings, and agriculture, as well as large-scale financial investments and major technological advances. Despite the increasingly clear economic potential of renewable energy and other clean technologies, existing policies and market momentum by themselves are likely to be insufficient to drive the energy transition fast enough. China's national carbon emissions trading system is a promising policy option, but some cost-effective emissions reductions are not subject to price signals. Strong policies and implementation plans are needed to support sustainable demand, improve energy efficiency, optimise the energy mix, enhance urban planning, and promote lifestyle changes. The COVID-19 pandemic brought about some changes and reduced overall carbon emissions in some areas, but these changes will require different strategies to address in the post-coronavirus era.

Lastly, we note that achieving carbon neutrality is, in itself, a major step towards sustainable development. Addressing climate change is one of the 17 sustainable development goals, and carbon neutrality is the advanced stage of addressing climate change. Achieving carbon neutrality can substantially contribute to progress on other sustainable development goals. A zero-carbon transition can significantly reduce the number of premature deaths caused by air pollution, which is what SDG #3, good health and well-being, aims to achieve. Carbon neutrality and sustainable development are inseparable and mutually reinforcing and should be planned, promoted, and implemented in parallel.

In light of what we have covered in this chapter, the following chapters of the book explore the nexus between carbon neutrality and sustainable development. We explore this nexus in the context of city clusters in China. We believe China's long-term plan of city cluster development is a major regional development initiative to move away from the megacity development pattern. The already shifting transitions and transformations are towards healthy urban–rural development directions, pathways to achieve low-carbon development and carbon neutrality, and opportunities for high-quality urbanisation aligned with harmonious development. In the next chapter, we introduce China's city cluster development and then delve into 11 of the 19 city cluster plans (see Chaps. 3–14). The book maps these cases based on their significance, industrial structure, historical development, sustainability plan/agenda, development status, etc. We will then conclude the book (see Chap. 15) with a summary of directions for China's cluster city development and carbon neutrality plan.

References

1. Chen L, Msigwa G, Yang M, Osman AI, Fawzy S, Rooney DW, Yap PS (2022) Strategies to achieve a carbon neutral society: a review. Environ Chem Lett 1–34
2. Cheshmehzangi A (2020) The analysis of global warming patterns from 1970s to 2010s. Atmos Clim Sci 10(3):392–404
3. Cheshmehzangi A, Chen H (2021a) China's sustainability transitions: low carbon and climate-resilient plan for carbon neutral 2060. Springer Nature

4. Cheshmehzangi A, Chen H (2021b) Decarbonised race and new destination in China. In: China's sustainability transitions. Springer, Singapore, pp 53–69
5. Cheshmehzangi A, Dawodu A (2019) The review of sustainable development goals (SDGs): People, perspective and planning. In: Sustainable urban development in the age of climate change. Palgrave Macmillan, Singapore, pp 133–156
6. Cheshmehzangi A, Li H, Yang X (2018) Low carbon and climate resilient urban development in China. ADB Report, developed for the National Development Reform Committee (NDRC), Beijing, China
7. Climate Action Tracker (2020) China going carbon neutral before 2060 would lower warming projections by around 0.2 to 0.3 °C. Available from: https://climateactiontracker.org/press/china-carbon-neutral-before-2060-would-lower-warming-projections-by-around-2-to-3-tenths-of-a-degree/
8. He J, Li Z, Zhang X, Wang H, Dong W, Du E, Chang S, Ou X, Guo S, Tian Z, Gu A, Teng F, Hu B, Yang X, Chen S, Yao M, Yuan Z, Zhou L, Zhao X, Li Y, Zhang D (2022) Towards carbon neutrality: a study on China's long-term low-carbon transition pathways and strategies. Environ Sci Ecotechnol 9:100134
9. Hepburn C, Qi Y, Stern N, Ward B, Xie C, Zenghelis D (2021) Towards carbon neutrality and China's 14th five-year plan: clean energy transition, sustainable urban development, and investment priorities. Environ Sci Ecotechnol 8:100130
10. Institute of Climate Change and Sustainable Development of Tsinghua University (ICCSD) et al. (2022). China's long-term low-carbon development strategies and pathways: comprehensive report. Springer, Singapore
11. Intergovernmental Panel on Climate Change (IPCC) (2022) climate change 2022: mitigation of climate change. Available from: https://www.ipcc.ch/report/ar6/wg3/
12. Lockwood M (2013) The political sustainability of climate policy: the case of the UK climate change act. Glob Environ Chang 23(5):1339–1348
13. Mallapaty S (2020) How China could be carbon neutral by mid-century. Nature 586(7830):482–484
14. Masson-Delmotte V, Zhai P, Pörtner HO, Roberts D, Skea J, Shukla PR, Pirani A, Moufouma-Okia W, Péan C, Pidcock R, Connors S (2018) Global warming of 1.5 °C. An IPCC Special Report on the impacts of global warming 1(5). Available: https://www.c40knowledgehub.org/s/article/IPCC-Special-Report-on-Global-Warming-of-1-5-C?language=en_US&gclid=Cj0KCQjwguGYBhDRARIsAHgRm4_1ykGNPaKwk0uyCbaJ_P56IT45Q6DNRk7HNp_P_xiOJCfffeloPo4aAn0OEALw_wcB
15. Ministry of Foreign Affairs of China (2016a) China's position paper on the implementation of the 2030 agenda for sustainable development. Available from: https://www.fmprc.gov.cn/mfa_eng/topics_665678/2030kcxfzyc/201811/t20181129_701773.html
16. Ministry of Foreign Affairs of China (2016b) China's national plan on implementation of the 2030 agenda for sustainable development. Available from: https://www.fmprc.gov.cn/mfa_eng/topics_665678/2030kcxfzyc/201610/P020210525474814580752.pdf
17. Ministry of Foreign Affairs of China (2017) China's progress report on implementation of the 2030 agenda for sustainable development. Available from: https://www.fmprc.gov.cn/mfa_eng/topics_665678/2030kcxfzyc/201708/P020210525474833457038.pdf
18. Ministry of Foreign Affairs of China (2019) China's progress report on implementation of the 2030 agenda for sustainable development. Available from: https://www.fmprc.gov.cn/mfa_eng/topics_665678/2030kcxfzyc/201909/P020210525474868189879.pdf
19. Ministry of Foreign Affairs of China (2021a) China's VNR report on implementation of the 2030 agenda for sustainable development. Available from: https://www.fmprc.gov.cn/mfa_eng/topics_665678/2030kcxfzyc/202107/P020210912747378573318.pdf
20. Ministry of Foreign Affairs of China (2021b) China's progress report on implementation of the 2030 agenda for sustainable development. Available from: https://www.fmprc.gov.cn/mfa_eng/topics_665678/2030kcxfzyc/202109/P020211019152754484797.pdf

21. Ministry of Science and Technology of China (2002) Outline of science and technology for sustainable development 可持续发展科技纲要 (2001–2010). Available from: http://www.gov.cn/gongbao/content/2003/content_62625.htm
22. Ministry of Science and Technology of China (2022) Sci-tech action plan to support the goal of peaking carbon dioxide emissions and carbon neutrality 科技支撑碳达峰碳中和实施方案 (2022—2030). Available from: http://www.gov.cn/zhengce/zhengceku/2022-08/18/5705865/files/94318119b8464e2583a3d4284df9c855.pdf
23. National Development and Reform Commission of China (NDRC) (2021a) Working guidance for carbon dioxide peaking and carbon neutrality in full and faithful implementation of the new development philosophy 关于完整准确全面贯彻新发展理念做好碳达峰碳中和工作的意见. Available from: https://en.ndrc.gov.cn/policies/202110/t20211024_1300725.html
24. National Development and Reform Commission of China (NDRC) (2021b) Action plan for carbon dioxidei peaking before 2030 2030年前碳达峰行动方案. Available from: https://en.ndrc.gov.cn/policies/202110/t20211027_1301020.html
25. Shan Y, Guan D, Hubacek K, Zheng B, Davis SJ, Jia L, Liu J, Liu Z, Fromer N, Mi Z, Meng J (2018) City-level climate change mitigation in China. Sci Adv 4(6):eaaq0390
26. Shan Y, Huang Q, Guan D, Hubacek K (2020) China CO_2 emission accounts 2016–2017. Sci Data 7, 1–9
27. Guan Y, Shan Y, Huang Q, Chen H, Wang D, Hubacek K (2021) Assessment to China's recent emission pattern shifts. Earth's Future 9(11):e2021EF002241
28. Shi Q, Zheng B, Zheng Y, Tong D, Liu Y, Ma H, Hong C, Geng G, Guan D, He K, Zhang Q (2022) Co-benefits of CO_2 emission reduction from China's clean air actions between 2013–2020. Nat Commun 13(1):1–8
29. State Council of China (1994) China's agenda 21: white paper on China's population, environment, and development in the 21st century 中国21世纪议程:中国21世纪人口、环境与发展白皮书: Adopted at the 16th Executive Meeting of the State Council of the People's Republic of China on 25 March 1994
30. State Council of China (2003) Program of action for sustainable development in China in the early 21st century 中国21世纪初可持续发展行动纲要. Available from: http://www.gov.cn/gongbao/content/2003/content_62606.htm
31. State Council of China (2007) Suggestions on promoting the sustainable development of resource-based cities 关于促进资源型城市可持续发展的若干意见. Available from: http://www.gov.cn/zhengce/content/2008-03/28/content_4941.htm
32. State Council of China (2013) National plan for sustainable development of resource-based cities of China 全国资源型城市可持续发展规划 (2013–2020年). Available from: http://www.gov.cn/zhengce/content/2013-12/02/content_4549.htm
33. State Council of China (2016) Development plan for China's innovation demonstration zones in response to the implementation of the 2030 agenda for sustainable development. 中国落实2030年可持续发展议程创新示范区建设方案. Available from: http://www.gov.cn/zhengce/content/2016-12/13/content_5147412.htm
34. United Nations (UN) (2016) Transforming our world: the 2030 agenda for sustainable development. Available from: https://sdgs.un.org/2030agenda
35. Wei YM, Chen K, Kang JN, Chen W, Wang XY, Zhang X (2022) Policy and management of carbon peaking and carbon neutrality: a literature review. Engineering
36. World Resources Institute (WRI) (2020) Accelerating the net-zero transition: strategic action for China's 14th five-year plan. Available from: https://www.wri.org/research/accelerating-net-zero-transition-strategic-action-chinas-14th-five-year-plan
37. Yang H, Huang X, Hu J, Thompson JR, Flower RJ (2022) Achievements, challenges and global implications of China's carbon neutral pledge. Front Environ Sci Eng 16(8):1–4
38. Zhou M, Hu T (2021) Analysis of carbon emission status under the carbon neutral target in China for Earth's atmospheric balance. In: IOP conference series: earth and environmental science, vol 804, No 4, p 042082. IOP Publishing

39. Zhao X, Ma X, Chen B, Shang Y, Song M (2022) Challenges toward carbon neutrality in China: strategies and countermeasures. Resour Conserv Recycl 176:105959
40. Zhuo Z, Du E, Zhang N, Nielsen CP, Lu X, Xiao J, Wu J, Kang C (2022) Cost increase in the electricity supply to achieve carbon neutrality in China. Nat Commun 13(1):1–13

Chapter 2
China's City Cluster Development: From Megacities Development Trend to Urban Agglomeration Plan

2.1 City Cluster Development: An Urban-Led Development, Regional Development, Economic Growth and Development, or All?

As mentioned in the introduction chapter, the focus of the book is on the context of city cluster development in China. The content of the study is the exploration of the nexus between sustainable development and carbon neutrality. Thus, we hope the book provides enough evidence of what is happing and what is expected to happen in the coming years and decades. China's carbon neutrality plan is a relatively long-term of almost four decades. Thus, it requires careful planning, innovation, technological integration, and management. Undoubtedly, there are evident bottlenecks down the line and each region has to make efforts in addressing their problems, strategise new plans and solutions, and develop holistically and healthily to achieve the regional indicators, targets, and goals. We also aim to provide a summary of reports, plans, pathways, directions, and strategies of important regions in China. Thus, we assert the book could serve as a mapping study of city clusters in China, based on their provincial and local reports, national plans, and important aspects of transportation and communication links.

It is not only China that has such an approach to its forthcoming development plans. Other countries and regions also have adopted and implemented a similar approach of city cluster development or regional development. A few examples include—but not limited to - Australia, India, Japan, the European Union, the United States, etc. According to the Asian Development Bank (ADB) 's report, city cluster development is an urban-led development strategy [5], p. 111, which is critical for regional development and the provision of infrastructure and services:

> City cluster development takes into account the provision of infrastructure and services in connection with spark potentials of economic growth and dispenses with the urban-rural dichotomy of traditional development planning, recognizing that urban centers are not only hubs for economic growth but also service centers for surrounding areas.

A. Cheshmehzangi and T. Tang, *China's City Cluster Development in the Race to Carbon Neutrality*, Urban Sustainability, https://doi.org/10.1007/978-981-19-7673-5_2

Associated with urban and regional planning fields, city cluster development "*is composed of the economic and social influences emanating from a particular city*" [7]. Other scholars have also described the term in a similar way, mainly to understand the implications of regional development, rural–urban development, economic growth and development, and more than just a centralised approach to urbanisation. The many descriptions of the term indicate having an urban core or multiple cores and a territorial or regional development of the core/cores. Thus, we see a sort of development network heavily relying on enhancing urban systems, urban–rural cooperation, and the provision of infrastructure and services to the region and between the regions.

Many countries globally suffer from polarised or centralised urbanisation patterns. For instance, some countries have a sort of imbalanced urbanisation and growth pattern, leading to disparities between regions. For instance, records show that more than one-third of England's population (excluding other states of the United Kingdom) live in the capital city of London. A similar trend is seen elsewhere in both developed and developing countries, particularly in the smaller nations. Several years ago, Sri Lanka had the same plan for Colombo, developing it into a vast megapolis stretched at one side of the country. Back in 2006, it was aimed to turn Colombo into South Asia's transshipment and logistic center. Another example is the capital city of Manila in the Philippines, which has already grown to be a rather daunting example of centralised urbanisation. However, such a development model could create a polarised effect across a country and its regions, promote disparities in development patterns, and lead to longer-term economic growth and development issues. If ignored, they could sabotage national planning frameworks and create imbalanced urbanisation. Yet, our book focuses not on such debates but on how city cluster development could help reverse such developmental trends or patterns. Thus, city cluster development could be seen as a regional development model, such as creating urban networks, urban corridors, industrial clusters, etc.

Urban corridors and networks are also known as a sort of regional development plan in both contemporary and ancient times. In the US, for example, we have urban corridors that include several large cities and regions. They also have the ongoing development plan of "America 2050", also regarded as the 'Emerging Megaregion' development, which includes several major regional development projects at both state- and cross-state levels. Elsewhere, in South Africa for example, we have a row of major cities in the coastal regions. They also proposed a new radical plan for smaller towns in the country, through which the national spatial development framework considers a holistic regional development that includes both inland and coastal areas. In Australia, we have concentrated populations in particular regions in the east. In Japan, we have city groups and urban networks that help enhance infrastructure, transportation, and communications development and provision. In Chile, and due to the country's geographical characteristics, major cities and regions are connected in a corridor to the west. In Iran, cities at the northern coasts alongside the Caspian Sea and the southern coasts alongside the Persian Gulf are connected in corridors. In Andalusia, southern Spain, the same pattern exists and even connects with Morocco to some extent. In Italy, the northern region's cities and towns have long been connected in their confined—and yet interconnectedclusters. Subsequently, we are certain that similar cases or patterns of development can be identified and mapped easily if we had to explore each country's profile and development patterns.

Several years ago, when we held a workshop in Uganda, we noted the significant difference between having the capital city of Kampala growing towards a 10–15 million populated city (i.e., a megacity) and having five clusters or regions of 2–3 million people across the country. A year after (in 2014), the central government proposed a plan that considered the idea of multiple regions rather than having another Lagos (of Nigeria) or another Cairo (in Egypt) in Uganda. If that plan becomes a success, then Uganda should have several international airports, a better network of transportation and communications, a robust planning agenda, national environmental protection plans, etc. If they fail, then the situation may not only be unchanged but worsened. This example is just one of the many that help us realise the importance of regional interventions, particularly where urbanisation is imbalanced, polarised, or unsustainable. We continue to fight for the fact that urban development alone has its own problems. It lacks a considerate understanding of rural areas and townships and puts significant pressure on how rural–urban relations are sustained. The hallowed towns around Barcelona, Spain, remind us of the siphoning effect at the regional level. The deprived rural industrial hub in the US shows that megacity formation could be harmful in the long run. The growing poverty and disparity in many places show the widening gaps in societies around the world. At the same time, we note regional development could become more comprehensive and could enhance cooperation, coordination, and integration between cities, towns, villages, and remote areas of a region. And city cluster development is an example that is worth further exploration and study. Thus, in this book, we aim to delve into the context of city clusters more than any time before, looking at how it could lead to sustainable development pathways and ultimately a carbon–neutral development (if to be taken forward in the right way).

We note that throughout the book, we intentionally and exchangeably refer to the term 'city cluster' as other commonly-used terms as well, such as 'urban agglomeration', 'region', 'megalopolis', 'urban network', etc. In essence, a city cluster or urban agglomeration is a type of regional development. Thus, aligned with urbanisation and urban development patterns of China, city cluster development is a crucial study area. From both policy and practice perspectives, there is a lot to be considered at the regional or city cluster level. They also play a significant role in achieving the national-level plans, yet through the localisation and implementation at the smaller and more manageable spatial scales. Next, we provide a brief summary of China's city cluster development plan and then summarise the structure of the book.

2.2 A Brief Summary of China's City Cluster Development Plan

Aligned with China's new-type urbanisation plan [1, 2] and high-quality urbanisation targets, (China's) city cluster development is seen as the backbone of what could be regarded as sustainable development for achieving the 2030 carbon peak and the 2060 carbon neutrality plan. The city cluster development is particularly important as a comprehensive plan recognised to change previous and ongoing development trends in various regions in China. According to Dezan Shira and Associates [6], China's city cluster development is recognised as the evolution of China's urban planning:

> With an urban population on course to hit 1 billion by 2030, the country plans to build 19 city clusters to sustainably manage its rapid urbanisation. China's city cluster plan consists of 8 medium and 8 small sized clusters as well as 3 "super" clusters. The central government prioritises these 3 super clusters to become world-class clusters by 2020, meaning that they will be the most innovative and internationally competitive, driving national economic development.

As it has been seen, the three super clusters of the Pearl River Delta, the Yangtze River Delta, and the Beijing-Tianjin-Hebei cluster have already proven to be leading regions and city clusters at the national and international levels. Not only that, but they also serve as the country's economic growth image. The ambitious development patterns that cost a lot of time and energy and require significant investment prove to be moving in the right direction to date. With increasing productivity, GDP increase, poverty reduction, and infrastructural development, this approach to regional development provide potential opportunities for cooperation, business and trade, expansion, and better planning. So far, the plan is to have 19 nationwide proposals of megalopolis (or in Chinese pinyin: Chéngshì Qún) development, which are aimed to enhance regional and local development through communication and transportation links. If successful, it is believed the number of city clusters could perhaps increase—which is expected to be 22 in the near future.

Nonetheless, as China's go-west urbanisation plan has not yet succeeded as initially anticipated, there seem to be ongoing issues with regard to sustaining a healthy urbanisation process beyond just the creation of megacities and now megaregions. Thus, the high-quality urbanisation plan is proposed to make some transformations following up on the earlier proposal of the new-type urbanisation plan (NUP) in 2014. To reduce the regional disparities, the country has a long way to go. But despite the concerns, "*China has a record for historical development achievements, and achieving these records in its own unique way*" [4]. In a way, the city cluster development plan is a designation by the Chinese government to enhance the development and growth of regions more than just larger-scale megacities. At first, the number of city clusters was limited to three super clusters in 2000 and then 13 megalopolises across the country. The shift from three to 13 was based on the development of new city clusters based on their regional importance. In 2012, as shown in the Economist Intelligence Unit (EIU) reports, the move was towards creating 13 city clusters, which were proposed to be achieved by 2020 (EIU 2012). These city clusters or megalopolises were: (1) Greater Beijing (Jing-Jin-Ji), (2) Greater Changsha (Chan-Zu-Tan), (3) Greater Shanghai (including Suzhou), (4) Greater Xi'an, (5) Chongqing, (6) Chengdu, (7) Greater Zhengzhou, (8) Greater Guangzhou, (9) Hefei Economic Circle (Including Lu'an, Huainan, and Chaohu), (10) Greater Shenyang, (11) the Shandong Peninsula, (12) Shenzhen, and (13) Wuhan. In other reports, however, there are indications that only nine of the city clusters were officially approved by 2018. Almost half way through the emergence of the NUP, it was in 2017 when the National Development and Reform Commission (NDRC) eventually proposed the nationwide city cluster development plan of 19 city clusters. The report then suggested completing six city clusters in 2016, five in 2017, and eight more new ones by 2020. As it stands now (August 2022), China has already achieved the development of all these 19 city clusters and continues to work on its regional development plans. In these transitional phases, the city clusters were renamed, and some were coupled (such as Chongqing and Chengdu megalopolises). The intention was to avoid naming the city clusters based on one city (such as Greater Shanghai, Greater Beijing, Greater Changsha, etc.) and move towards regional recognition. The list of all 19 city clusters is shown below:

2.3 Super Clusters

1. Beijing-Tianjin-Hebei, mainly located in three locations near the capital city of Beijing and including Beijing, north-east of China;
2. Yangtze River Delta, comprised of several major cities around Shanghai and including Shanghai, eastern coastal areas of China;
3. Pearl River Delta, mainly located in Guangdong Province, south of China.

2.4 Medium and Small Clusters—8 in Each Category

1. Shandong Byland or Shandong Peninsula, comprised of three economic circles or clusters in Shandong Province, north-east of China;
2. Middle Reaches of the Yangtze River or the Yangtze River Midstream, mainly as part of the inland and middle areas of the Yangtze River, located in central China;
3. Guangdong-Fujian-Zhejiang Coastal Region, covering an stretch of coastal areas in between three provinces of Guangdong, Fujian, and Zhejiang, in south, south-east, and east of China;
4. Chengdu-Chongqing, located between Chongqing and Sichuan Province, in inland areas of the western regions;
5. Guanzhong Plains, covering mainly central areas around Xi'An;

6. Central Plains, covering mainly central areas between the Greater Beijing and the Greater Shanghai regions;
7. Beibu Gulf/Bay, or Guangxi-Hainan, covering areas between Guangxi and Hainan Provinces, south of China;
8. Central Yunnan, covering only parts of the Yunnan Province, south of China;
9. Harbin-Changchun, covering areas between two provinces in north-east of China;
10. Central-and-Southern Liaoning, neighbouring the Korean Peninsula in north-east of China;
11. Lanzhou-Xining, located in western inland areas in China;
12. Ningxia Section of yellow River, close to the previous cluster, and located in north-western regions of China;
13. Central Shanxi, covering an area in between the Beijing-Tianjin-Hebei cluster and Central plains;
14. Hohhot-Baotou-Ordos-Yulin, located in north inland areas of China;
15. Central Guizhou, mainly covering areas in and around Guiyang, south of China;
16. Northern Slope of Tianshan Mountains, covering regions in Xinjang Province, north-west of China.

As shown in Fig. 2.1, the 19 approved city clusters were promoted in the 14th Five-Year-Plan (FYP) 's proposed plans until the year 2025 (see Fig. 2.3 for the overall plan). More importantly, it is evident that China's city cluster development is aligned with the national-level spatial pattern of urbanisation, which was approved prior to the announcement of all 19 city clusters. As shown in Fig. 2.2, all 19 city clusters are located in the three horizontal and two vertical urbanisation strategic patterns in mainland China. In this spatial pattern, there are allocated regional development plans, transportation networks, eco-environmental protection areas, economic hubs and belts, coordinated regional development plans, and deepening reforms and opening-up strategies. Figure 2.3 summarises the outline of the 14th Five-Year Plan (2021–2025) for National Economic and Social Development and Vision 2035. This plan is ongoing and is the first five-year plan after the approval of all 19 city clusters and the pledge to carbon neutrality. It is, therefore, the foundation of this major pathway until the year 2060.

2.5 China's City Cluster Development and Carbon Neutrality

To date, no published scholarly work has explored China's carbon neutrality in the context of its city clusters. This book aims to fill this gap by mapping China's city clusters in the race to carbon neutrality. It is important to note that available information is limited, and there is little information in the existing literature. We hope this could be a major field of research where regional development and city cluster development could be looked at more carefully and aligned with sustainable development

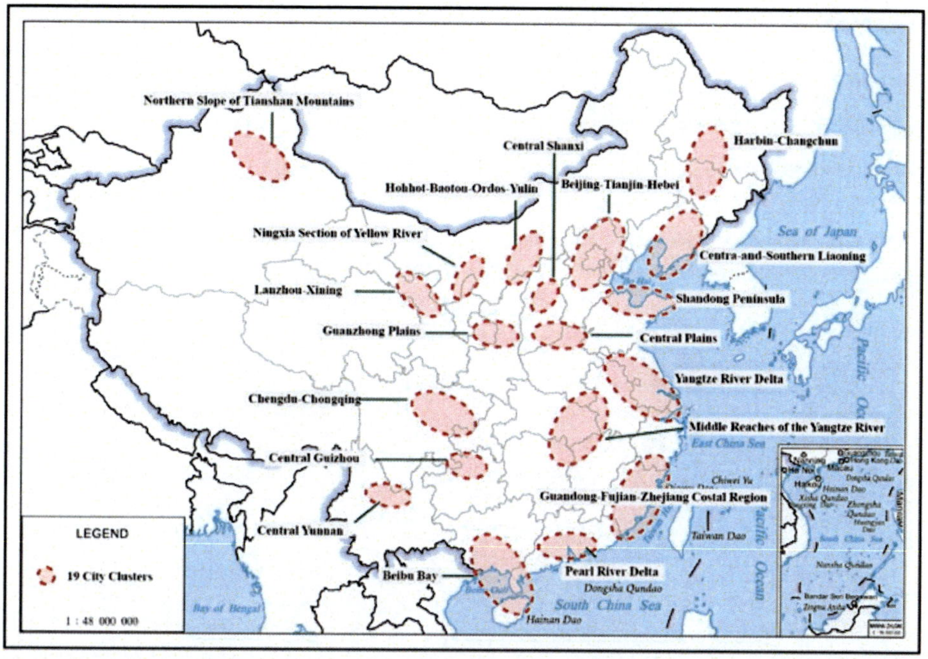

Fig. 2.1 China's 19 city clusters or urban agglomerations in the 14th Five-Year Plan. *Source* Drawn by the authors and adapted from the Chinese Government sources

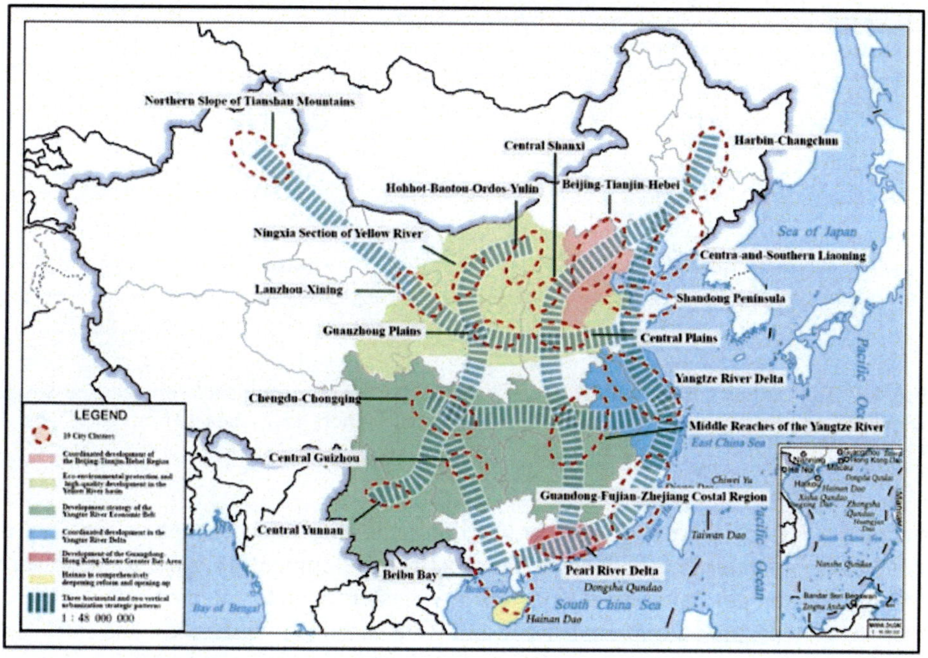

Fig. 2.2 China's spatial pattern of urbanisation in the 14th Five-Year Plan. *Source* Drawn by the authors and adapted from the Chinese Government sources

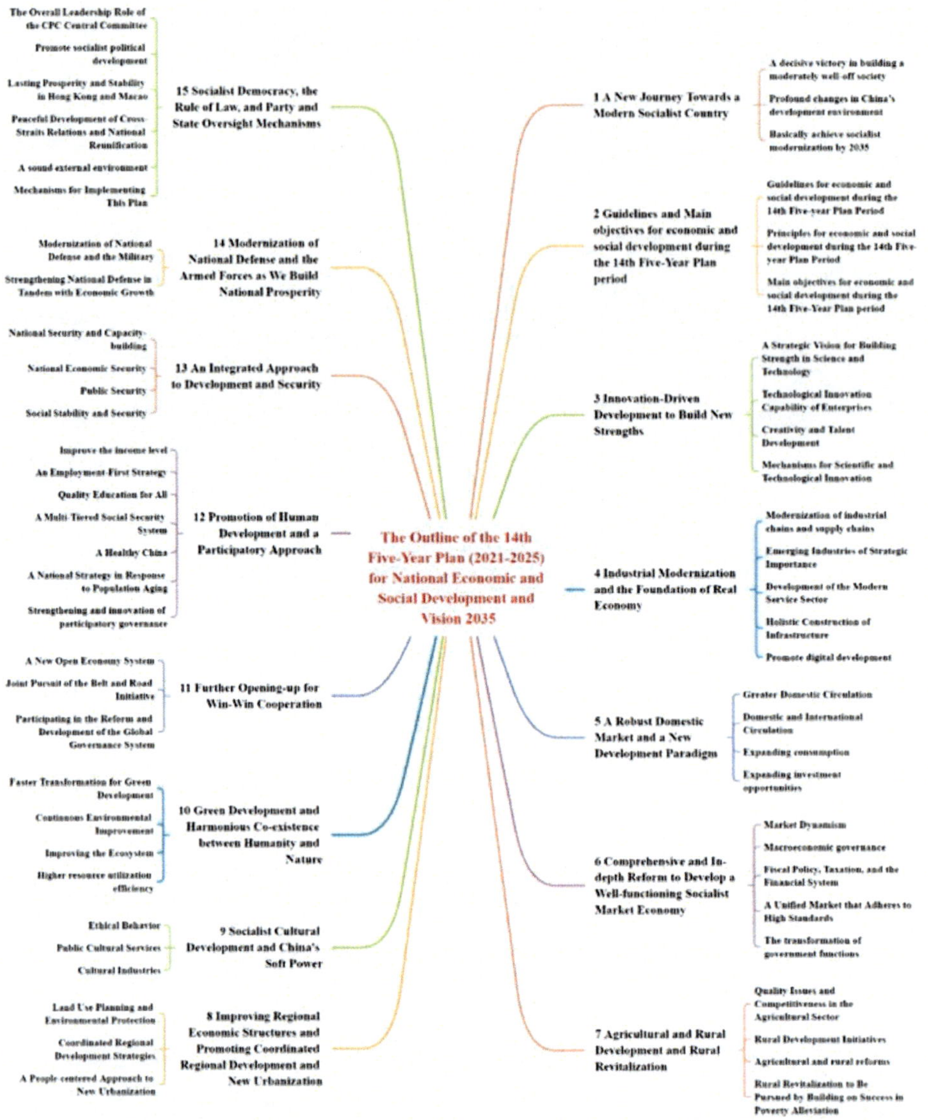

Fig. 2.3 The Outline of the 14th Five-Year Plan (2021–2025) for National Economic and Social Development and Vision 2035. *Source* Drawn by the authors and adapted from the Chinese Government sources

pathways, strategies, and directions. Carbon neutrality is an excuse or reason—if not a path—to achieve the ideals of sustainable development. In order to make sure China is carbon neutral down the line, its city clusters have to adopt policies, contextualise agendas, plan comprehensively, innovate sustainable practices, and integrate technologies that are context-specific, revolutionary (or game-changing), and zero-carbon. The path to the 2030 carbon peak targets and the 2060 carbon neutral plan would not be easy for the most populated country and now the second strongest economy worldwide. As urbanisation and industrialisation are still ongoing with the country's economic growth and development, China's pledge to carbon neutrality would mean a combination of transitional steps and transformational directions like none seen before.

We note again that information about the topic of carbon neutrality is relatively new or scarce. Only a few months after China's pledge to carbon neutrality, we put together a book about this new topic, titled '*China's Sustainability Transitions: Low Carbon and Climate-Resilient Plan for Carbon Neutral 2060*' [3]. It was the first published book related to China's carbon neutrality plan, and it received a high level of attention from scholars in China and internationally. Following that success, we recognised the importance of how the plan is expected to be implemented and at what scale it is important to see such implementation. Then, the idea of analysing city clusters became a suitable way that helped us evaluate what China's plan is and how it would be directed.

We confirm much of the information provided in this book is from valid and reliable resources, mainly from governmental reports and documents, transcripts of provincial and municipal level reports, openly accessible documents related to city clusters, and reports by practitioners or relevant stakeholders and media. Some of these reports are translated, and some are evaluated to highlight each of the city clusters' plans for carbon neutrality. However, only 11 out of 19 city clusters have enough information for us to provide a good picture of the conditions now and the future plans ahead. For the remaining eight city clusters, there is either little/no information available or there is no plan for carbon neutrality to date. The race to carbon neutrality, as the book title entails, is still ongoing and yet to be even more stimulating.

2.6 Structure and Summary of Case Study Chapters

This book maps 11 city clusters in the race to China's national-level carbon neutrality plan. The nexus between sustainable development, the recent and ongoing progress, and future sustainability plans are highlighted very carefully in each of the following 11 case study chapters. We explicitly focus on how the current and future plans are associated with the national-level carbon neutrality plan. We note there are some evident regional disparities and similarities, which we hope could be highlighted and understood through this mapping study of city cluster cases.

At first, we explore the three super clusters before looking at eight other clusters. These case study chapters are Chaps. 3–13, each dedicated to individual city clusters (11 out of 19). In Chaps. 3–5, we first explore the three super clusters of the Beijing-Tianjin-Hebei city, the Yangtze River Delta, and the Pearl River Delta city. Chapter 6 covers the Chengdu-Chongqing city cluster, which is known as an important historical economic zone with major plans for western China. In Chap. 7, we explore the Yangtze Mid-River Delta, or the middle reaches of the Yangtze River, which is a dynamic growth pole in the form of a middle triangle in inland areas of the

Yangtze River. Chapter 8 covers the Shandong Byland or the Shandong Peninsula city cluster, formed from three economic circles and industrial clusters. In Chap. 9, we delve in the case of the Guangdong-Fujian-Zhejian coastal region, one of the largest city clusters in China shared between three coastal provinces as a network corridor. Chapter 10 is focused on a rising inland city cluster, namely the Guanzhong Plain city cluster, where we explore the complexity of a complete industrial system. Chapter 11 maps the Central Plains city cluster where green transformation is visible and developing fast. Chapter 12 covers the Beibu Gulf city cluster between Guangxi and Hainan Provinces of the south, recognised as a new strategic pivot to foster new growth drivers in the region and across the borders. Lastly, in Chap. 13, we delve into the Central Yunnan city cluster, an important cluster for regional industrial development, tourism, and sustainable production. After the following 11 case study chapters, we conclude the book through brief discussions and summaries of China's city cluster development and its future directions. In particular, our discussions and conclusions are associated with China's carbon neutrality plan and sustainable development pathways. We hope the following chapters are informative, comprehensive, and a good platform for knowledge sharing for researchers, practitioners, educators, students, and governmental officials.

References

1. Cheshmehzangi A (2014) The urban and urbanism: China's new urbanization and approaches towards comprehensive development. Int J Interdiscip Environ Stud 8(3–4):1–12
2. Cheshmehzangi A (2016) China's New-Type Urbanisation Plan (NUP) and the foreseeing challenges for decarbonization of cities: a review. Energy Procedia 104:146–152
3. Cheshmehzangi A, Chen H (2021) China's sustainability transitions: low carbon and climate-resilient plan for carbon neutral 2060. Springer, Singapore
4. China Briefing (2018) China's city clusters: the plan to develop 19 super-regions. Available from: https://www.china-briefing.com/news/chinas-city-clusters-plan-to-transform-into-19-super-regions/
5. Choe K, Laquian A (2008) City cluster development: toward an urban-led development strategy for Asia, ADB Report. Asian Development Bank (ADB), Manila, p 111. Available from: https://www.adb.org/sites/default/files/publication/27555/city-cluster-development.pdf
6. Dezan Shira and Associates (2018) China's city cluster plan. Available from: https://www.dezshira.com/multimedia/view/china-s-city-cluster-plan-7849
7. Friedmann J (1992) Empowerment: the politics of alternative development. Blackwell, Cambridge, MA

Chapter 3
Beijing-Tianjin-Hebei City Cluster: From a Political Center to a Major Industrial and Innovation Chain

Check for updates

3.1 A General Introduction to the Beijing-Tianjin-Hebei City Cluster

The concept of the Beijing-Tianjin-Hebei city cluster developed from the capital city agglomeration. The city cluster includes cities and areas from three provinces. These include Beijing and Tianjin municipalities, Baoding, Tangshan, Langfang, Shijiazhuang, Qinhuangdao, Zhangjiakou, Chengde, Cangzhou, Hengshui, Xingtai, Handan, and Anyang in Henan Province. Among them, the central core functional areas are Beijing, Tianjin, Baoding, and Langfang. The Beijing-Tianjin-Hebei city cluster has a population of 115.44 million [2], covering an area of 218,000 square kilometers. It accounts for 2.3% of China's land area. By 2020, the GDP reached 8.6 trillion RMB or 8.5% of the country's GDP. It is estimated that the per capita GDP is approximately 77,000 RMB. In addition, the major project of the Beijing-Tianjin-Hebei Integration plan (also known as Jing-Jin-Ji) is the development of a capital economic circle [47] that has a significant impact on local economies, regional industries [39], and socio-political development.

In the Beijing-Tianjin-Hebei city cluster, Beijing is the national capital and the center of the city cluster. The level of medical care, educational resources, and employment opportunities in Beijing are higher than those in neighboring cities and other cities in China, so the siphoning effect of Beijing on the population is very strong [49, 62]. In recent years, the population of the Beijing-Tianjin-Hebei city cluster and even the whole country's population has constantly been flocking to Beijing [8, 10, 29]. In this region, the main outflow of the population is Hebei, and population inflow is mainly from Beijing. As Beijing's population continues to expand, it impacts the Beijing-Tianjin-Hebei region's economic development, which is seriously uncoordinated [51]. For decades, the focus in the region has been on the development of Beijing and Tianjin, two stronghold cities, and this has caused an imbalance in this city cluster (i.e., through uncoordinated-urbanisation hindered development). The subsequent results are that the development of the Hebei region is relatively weak. Therefore, the "Beijing-Tianjin-Hebei Coordinated Development

A. Cheshmehzangi and T. Tang, *China's City Cluster Development in the Race to Carbon Neutrality*, Urban Sustainability, https://doi.org/10.1007/978-981-19-7673-5_3

Planning Outline" proposed to control the population of Beijing [48], "*strictly control the increment, ease the stock, combined with the control of the population size in Beijing*" [2]. In this regard, a major recent change is to create better coordination and balance at the regional level, which is essential for stepping up toward further developments aligned with the carbon neutrality plan.

3.2 A Historical Analysis of the Beijing-Tianjin-Hebei City Cluster

The historical retrospect of the cluster suggests the 'Origin of Tale of Two Cities'. The development pattern along the Taihang Piedmont Plain goes before the Sui and Tang Dynasties. In the last two decades, there have been major spatial and environmental factors related to heavy material distribution [54], management of groundwater resources, agricultural irrigation [40], etc. In prehistoric times, the Beijing-Tianjin-Hebei region was one of the critical regions for ancient human activities and reproduction in China [3]. From the population-land-industry perspective of contemporary times, similar essential factors are still in place.

The basins of the Taihang Mountains, where Beijing-Tianjin-Hebei is located, provided a good environment for ancient human life and hunting. Sanggan River is an important region for ancient human reproduction in the Paleolithic Age, Nihewan, from more than 200 million years ago to more than 3,000 BC [55]. More than 130 Neolithic sites are also extremely rare worldwide [3]. Forming such a rich cultural heritage of the early human superimposed cultural layer plays a significant role in the later (and more recent) development of the region. So far, the earliest heritage site was discovered in the transition zone between buried hills and plains in Peking Man in Zhoukoudian, where people lived about 70,000 to 200,000 years ago. They then migrated again due to changes in extreme environmental conditions. The experts also identified that about 120,000 to 125,000 years ago, the Xujiayao people selected this region as an essential gathering place for the westward migration of Peak Cave people in Beijing [3]. In the New Era, about 10,000 to 5,000 BC, more ancient human settlements appeared in front of the Taihang Mountains and the North China Plain. During this period, a large number of pottery jars for storing grain and food appeared, which opened the prelude to transforming human nature in the plain area of North China [36, 37]. In the middle of the new era, North China's grain crops were planted with millet, rice, millet, beans, wheat, and so on [1, 23]. Therefore, from the development of prehistoric civilization and the space succession, the Taihang piedmont plain and the western mountainous area have a close cultural origin to the human settlements of the region [19, 33, 43]. The human activity areas gradually moved from the mountainous regions to the buried hill and the plain areas.

From the Warring States Period to Sui and Tang Dynasty, the population, economy, and cultural center of the Beijing-Tianjin-Hebei area were distributed in the area of 20 km before Taihang Mountain [12–14]. At the same time, military towns are distributed along the Sanggan River and the south of the Great Wall [41, 45]. In the Shang Dynasty, Xingtai became the capital of Zuyi and took the lead in developing it into a city [46, 50]. During the Western Zhou dynasty, the Yan State in the north and the Xing Kingdom in the south became the two vassal states in the Beijing-Tianjin-Hebei region [17, 27]. In the Warring States Period, Ji in Yan and Handan in Zhao were two major political, economic, military, and cultural centers in the region and maintained a relatively stable development pattern [57]. It is worth noting that the Yellow River estuary in the Bohai Bay. Hence, the North China Plain at this time, most of the region belongs to the water country, not suitable for agriculture and urban development [3]. Because the Yellow River changed its course frequently and frequently flooded in the past 3,000 years, it caused great harm to the North China Plain (ibid). Thus, the spatial distribution of the cities and towns in Northern China was stable for a long time, and there was no trend of large-scale eastward expansion. It was not until the Jin Dynasty that the Yellow River basically swayed south of Hebei Province, making the North China Plain more suitable for farming as a whole [3, 22]. At this time, with the prosperity of trade and transportation of the Beijing-Hangzhou Grand Canal, Tianjin and Cangzhou got into a stage of unprecedented development [3]. In addition, many tributaries of the Haihe River basin also promoted the prosperity of the city with the aid of water transport development [24], such as Hejian, Renqiu, and other cities.

3.3 An Overview of Industrial Structure in the Beijing-Tianjin-Hebei City Cluster

The Beijing-Tianjin-Hebei city cluster is China's political and economic center and an important core area in the north. The development of its industrial structure has a vital role in promoting China's national economy. Therefore, the development of the Beijing-Tianjin-Hebei region has become an important symbol [61] to measure regional competitiveness and development.

Here, we refer to the basic situation of the economy and industrial structure [32, 53] in the Beijing-Tianjin-Hebei city cluster. The Beijing-Tianjin-Hebei region is now transitioning from secondary to tertiary industry [6]. From 2009 to 2018, the scale of the tertiary sector in Beijing-Tianjin-Hebei continues to grow, and its proportion continued to expand, while the proportion of secondary industry continues to decline [3]. Meanwhile, there is little change in the primary sector. The inverted pyramid industry pattern of "321" (ibid) has basically taken shape.

In terms of GDP, the proportion and development trend of the three industries in the Beijing-Tianjin-Hebei region are the same as that of the whole country [18]. The proportion of the secondary sector in the gross domestic product is slightly higher than that of the national total. For the proportion of the labour force, the percentage of the primary industry in the Beijing-Tianjin-Hebei region is obviously lower than the national level, and the percentage in the secondary and tertiary industries is obviously higher than the country level [3]. In order to solve these problems, Beijing, Tianjin, and Hebei should give full play to their advantages, attract talents, strengthen the construction of an urban "soft environment," attach importance to the development of education level, science, technology, and culture in the region, and implement the upgrading and transformation of industrial structure.

The 14th Five-Year Plan proposes to promote the in-depth integration of the Beijing-Tianjin-Hebei industrial 'chain' and 'innovation chain'. The deep integration of the industrial chain and innovation chain is the key to solving the difficulty of landing Beijing's scientific and technological achievements in Tianjin and Hebei and the challenges of regional industrial coordination [3]. It is also an important measure to deal with the global industrial chain reconstruction and promote the construction of world-class industrial clusters in Beijing and Tianjin. Moreover, the Beijing-Tianjin-Hebei industrial chain and innovation chain integration process is also facing the following seven aspects (summarised and translated from reports by the [3] of the problem:

(1) ***The proportion of "strong point" sectors in the industrial chain that gain comparative advantage through technological innovation is low***
In terms of specific industry sectors, the proportion of technology-oriented industries in the top comparative advantage industries in the Beijing-Tianjin-Hebei city cluster is relatively low [7, 9, 38]. The number of new utility patents of enterprises in the Beijing-Tianjin-Hebei city cluster shows an inverted pyramid distribution, that is, among all the comparative advantage industries involved in the enterprises, the head of a few enterprises control nearly half of the new type utility patents. This reflects that the proportion of key nodes in the industrial chain of the Beijing-Tianjin-Hebei city cluster is still low, and most of the key nodes still maintain the relative competitiveness of the industry with non-technical advantages [26]. From the specific space carrier, Xingtai, Handan, Hengshui, and Chengde's comparative advantage industries of technological achievements stock are relatively low [5]. In these cities, the comparative advantage industries have less than one new utility patent per household. In the case of endogenous power shortage [34], the relative channels for these cities' comparative advantages industries to obtain technological achievements in the industry are not smooth enough.

(2) ***The industrial chain integrity of a small number of cities showed a downward trend, resulting in the original industrial chain within the industry distance elongated***
In the industrial chain of the Beijing-Tianjin-Hebei city cluster, the comparative advantage industry categories of other cities have increased, except for Shijiazhuang and Baoding [4]. From 2015 to 2019, the density of key nodes in Shijiazhuang and Baoding industrial chains in some regions has decreased, as evidenced by the elongated distance between the dominant industries in the industrial chain and the significant decline in the number of directly related dominant industries. In the evolution of the industrial chain in Shijiazhuang, the distance between the secondary industry sectors with comparative advantage

has been lengthened [5]. In the process of industrial chain evolution in Baoding, the distance between the secondary and tertiary industries shows an expanding trend. The disappearance of supporting industries has a particular impact on the development of the remaining comparative advantage industries.

(3) ***The lack of key nodes in the industrial chain of most cities in the Beijing-Tianjin-Hebei city cluster is not conducive to the development of the end comparative advantage industries***

In the process of industrial chain development in the Beijing-Tianjin-Hebei city cluster, the lack of some "branch chain" comparative advantage industries are apparent (also see [35]). The lack of a "bifurcation point" in the extension process of these industrial chains is not conducive to the long-term development of the comparative advantage industries at the end of the "branch chain". It then becomes the key "sticking point" affecting the whole industrial chain's competitiveness [3]. From the actual situation of the industrial chain evolution of the cities in the Beijing-Tianjin-Hebei city cluster from 2015 to 2019, the lack of hub industry in the early "branch chain" has not been effectively supplemented. Compared with solving the "bottleneck" problem of key technologies in the innovation chain, it is feasible to fill the missing industries in the industrial chain [21, 60]. The key is whether the parties can introduce these missing industry sectors and embed them in the industrial chain, forming complementary regional industry linkages, strengthening and enhancing the competitiveness of the original industrial chain.

(4) ***The node cities in the Beijing-Tianjin-Hebei city cluster are underdeveloped, and most of the cities have a high degree of deviation from the development of the industrial chain and the innovation chain***

Compared with the Yangtze River Delta and the Pearl Delta, the proportion of node cities in the complex network of industrial sectors and innovation sectors in the Beijing-Tianjin-Hebei city cluster is low [30]. Suppose the number of industrial linkages between a city and other cities in the complex network of industry sectors is much greater than (or much less than) the number that the city has industrial linkages with other cities within complex networks of innovation sectors. In that case, a certain extent can reflect the city's industrial chain development and innovation chain development degree of deviation [3]. Compared with the Yangtze River Delta and the Pearl River delta, the number of cities in the Beijing-Tianjin-Hebei city cluster with a higher deviation between industrial chain and innovation chain development is more [30], and there is still an ample space for improvement in the integrated development of industrial and innovation chains.

(5) ***There are significant regional differences between high-tech manufacturing and producer services in the Beijing-Tianjin-Hebei city cluster, and there is a high degree of separation between the two sectors***

The coupling and synergetic development level of high-tech manufacturing and producer services in the Beijing-Tianjin-Hebei city cluster present obvious gradient differentiation [11]. There are also spatio-temporal evolution characteristics ad influencing factors of the urban service industry in this region, which play a significant part in promoting the synergistic development of the industries [59]. As a result, the high-tech manufacturing industry and producer services coupling synergy between better cities and coupling synergy gap between poor cities gradually widened. From the perspective of high-tech industry segments, the reasons that restrict the improvement of coupling synergy mainly include the following aspects: First, the manufacturing sector is highly separated from the producer service sector in the new generation of the information technology industry in the Beijing-Tianjin-Hebei city cluster [30]. Also, Cangzhou lacks the support of technological achievements. Second, the main gathering area of the high-end equipment manufacturing industry in the Beijing-Tianjin-Hebei city cluster lacks the support of technological achievements [56] and the technological achievements in the high-end equipment manufacturing field in Beijing. Thus, the source of innovation does not flow to the surrounding areas. Third, in the Beijing-Tianjin-Hebei city cluster, there is a high degree of separation between the manufacturing and producer services sectors [42]. For instance, Hengshui is a large manufacturing sector, and its biomedical technology is relatively insufficient.

(6) ***There is a significant gap between innovation source and innovation destination, which restricts the improvement of technology output efficiency***

No matter in the aspect of investment in innovation and R&D, or in attracting various types of research and development design main body gathering, the innovation source of Beijing has apparent advantages over Tianjin and Hebei [63]. This situation forms a regional distribution pattern of "core and periphery" innovation resources [44]. The innovation achievement circulation channels from the innovation source (i.e., core) to the innovation hinterland (i.e., periphery) are not well developed. Intellectual property protection agencies and science and technology intermediaries are mainly concentrated on the supply side of innovation resources (i.e., in Beijing), less on the demand side (i.e., in Tianjin and Hebei).

(7) ***The Beijing-Tianjin-Hebei industrial chain cultivation process to build the lack of a clear lead management organisation***

Among the key areas for the coordinated development of Beijing-Tianjin-Hebei, industrial synergy has the function of "hematopoiesis" [35]. At present, the process of industrial coordinated development of the Beijing-Tianjin-Hebei city cluster is not robust. This is due to the lack of a leading management department to guide the docking and integration of the three industrial chains. The game behavior based on their interests cannot eliminate the "prisoner's dilemma" [3]. First, it faces the problem of profit distribution in the process of cross-regional integration of the industrial chain. The high-end part of the industrial chain is the most profitable sector in the value chain. Every region wants to do high-end, but not every region has the conditions to do the high-end sector of the industrial chain (ibid). This leads to the weak foundation of the region that should play its own advantages to do a good job in basic supporting industry sectors. However, a radical type (competitive) high-end development does not seek the advantage of docking cooperation opportunities. It also does not want to do a good job of the advantages of regional service matching [3]. Second, the three places in Beijing, Tianjin, and Hebei in the industrial chain docking process are facing the "who led who took the lead" problem (also see [15]. The governments and industrial departments of the three places want to have a confident voice in the development process of the industrial chain of Beijing, Tianjin, and Hebei [25]. If one of the three places is the dominant department, then the other two places will also show some negative behaviour in the later period/stage [3], which is not conducive to the integration of the development of the industry chain.

3.4 Sustainable Development Plan: The Beijing-Tianjin-Hebei City Cluster Industrial Chain and Innovation Chain

In this city cluster, integration and development still have great room for improvement and promotion. Compared with other city clusters, the cities in the Beijing-Tianjin-Hebei city cluster have a lower degree of association with other cities in complex networks of industrial sectors and innovation sectors. Therefore, a more holistic input–output analysis with complex network analysis [58] would be suitable for assessing the region's conditions and future development capacities. Also, cities have a big gap between the industrial and innovation sectors. From the spatial structure of the "core-node-edge", industrial sectors and innovation sectors in the Beijing-Tianjin-Hebei city cluster have formed a development pattern with Beijing as the core and Tianjin as the node [3]. The proportion of nodal cities in city clusters is low, and the overall structure is unbalanced [28]. Therefore, we can summarise that the Beijing-Tianjin-Hebei city cluster industrial chain and innovation chain integration development has an ample space for improvement.

3.5 Summary: Pathways Towards Achieving China's Carbon Neutrality Plan (CNP): Opportunities and Challenges

First, it is clear that the region needs to focus on the layout of the industrial chain and enhance the industry chain technology content [4]. The focus is expected to be on Beijing's high-precision industries and regional strategic emerging industries, i.e., the industrial chain layout of key industries, such as new generation information technology, new energy, energy conservation, environmental protection, artificial intelligence, and modern transportation [3]. The primary emphasis will be placed on the scientific and technological content of the industrial chain to improve the regional innovation capacity continuously. This approach should lead to the further development of new products and technologies with high added value and provide advanced technical support for the industry chain (see [26]. In return, it realises the extension of the critical industrial chain to high value-added links through scientific and technical innovation.

Second, Beijing, Tianjin, and Hebei should adhere to the development concept of the integration of regional industrial chain and innovation chain. This could also benefit the development of a cross-regional innovation system [31]. As the core city, Beijing should focus on playing is role as a leading role, accelerate the construction of Zhongguancun National Independent Innovation Demonstration Zone [16], establish an innovative science and technology park combining production, learning and research, and promote the development of surrounding areas through innovation spillover. As the node city, Tianjin should strengthen the docking with Beijing's scientific and technological innovation and high-end industries, promote the deep integration of the industrial chain and innovation chain, and actively integrate into the regional industrial collaborative development pattern. As a marginal city, the cities of Hebei should strengthen independent innovation research and development, promote the integration of the undertaking industry and their own industrial development, and achieve industrial upgrading [3]. At the same time, the plan should aim to promote the Beijing-Tianjin innovation achievements landing in Hebei, consolidate the foundation for the transformation of accomplishments, and improve the project landing rate.

Third, the plan should strengthen market construction and give full play to the prominent role of enterprises in innovation. It has to adhere to the main status and leading role of enterprises in the innovation process and adopt the development model of "government policy + leading enterprises + financing guarantee + industrial chain SMEs" to enhance the international competitiveness of SMEs [3]. In return, this approach should improve the industrial ecosystem. We encourage leading enterprises in the Beijing-Tianjin-Hebei city cluster to establish platforms for entrepreneurship and innovation [3] and provide services such as innovation, investment and financing, and incubation so as to further promote the development of innovative and technological enterprises and enhance the supply capacity of regional scientific and technological innovation from the source of the innovation chain. This approach should give full play to the leading enterprises in the region in the R&D platform, industrial chain and sales channels, and other resources to provide rapid transformation channels for scientific and technological achievements and achieve commercial application [20]. Thus, it is anticipated that regional development will focus more on the weak links of the industrial chain through the industry chain investment and other ways to promote the development of domestic and foreign leading enterprises landing. It should then give full play to the agglomeration effect of leading enterprises.

Fourth and lastly, it is vital to focus on coordinating multi-chain coordinated development, multi-angle boost industrial chain, and innovation chain (i.e., double chain) integration development [52]. In the process of integration of the industry chain and innovation chain, various activities related to it will have a very important impact on the final integration effect. Therefore, the Beijing-Tianjin-Hebei city cluster should promote the coordinated development of multiple chains such as the service chain, financial chain, policy chain, and talent chain in a planned way [3]. By building a structural system of multi-chain coordinated development, we can better realise the linkage of each chain and each link, enhance the overall value of the integration of the industrial chain and innovation chain, and promote high-quality economic development.

References

1. Anderson EN (1988) The food of China. Yale University Press, New Haven, Connecticut
2. Baike (2020) Beijing-Tianjin-Hebei 京津冀 Available from: https://baike.so.com/doc/5515205-5748546.html
3. Beijing-Tianjin-Hebei Blue Book (2021) Report on the Development of Beijing, Tianjin and Hebei: Integration of the Industrial Chain and the Innovation Chain 京津冀蓝皮书: 京津冀发展报告(2021) 产业链与创新链融合发展 Available from: https://www.pishu.cn/zxzx/xwdt/568724.shtml
4. Chai Y, Lin X, Wang D (2021) Industrial structure transformation and layout optimization of beijing-tianjin-hebei region under carbon emission constraints. Sustainability 13(2):643
5. Chang Y, Zhang Q (2020) Industrial transfer and spatial structure optimization of Beijing, Tianjin and Hebei province. Int J Des Nat Ecodynamics 15(4):593–602
6. Chen J, Yuan H, Tian X, Zhang Y, Shi F (2019) What determines the diversity of CO_2 emission patterns in the Beijing-Tianjin-Hebei region of China? An analysis focusing on industrial structure change. J Clean Prod 228:1088–1098
7. Cheshmehzangi A (2016) China's New-Type Urbanisation Plan (NUP) and the foreseeing challenges for decarbonization of cities: a review. Energy Procedia 104:146–152
8. Cheshmehzangi A (2020) Towards urban energy management: a brief review of China's recent transition in urban energy demand. Low Carbon Econ 11(01):1
9. Cheshmehzangi A, Chen H (2021). Key suggestions and steps ahead for China's carbon neutrality plan. In: China's sustainability transitions. Springer, Singapore, pp 189–206
10. Cheshmehzangi A, Dawodu A, Sharifi A (2021) Sustainable Urbanism in China. Routledge, New York
11. Cong H, Zou D (2017) The research on the mechanism and spatial–temporal differentiation of the coupling coordination development based on industrial cluster agglomeration. Clust Comput 20(1):195–213
12. Du L, Peng X, Zhao Y, Lyu J, Pan M (2021) From nature-based to engineering-based: the interaction process and turning point of the city-water system relationship in the North China plain. Appl Geogr 135:102556
13. Du L, Wang F, Zhang Q, Fang Y (2022) Adaptation of urban distribution to the fluvial geomorphic environment and the reconstruction of the river system structure by urban distribution in Haihe Plain in the past 2,200 years. River Res Appl. Available from: https://onlinelibrary.wiley.com/doi/full/https://doi.org/10.1002/rra.3935
14. Fu C, Cao W (2019) Introduction to the urban history of China. Springer, Cham
15. Gao S, Wang H (2022) Analysis of industrial carbon transfer in Beijing-Tianjin-Hebei city cluster and surrounding areas. Front Environ Sci 10. Available from: https://www.frontiersin.org/articles/https://doi.org/10.3389/fenvs.2022.895142/full
16. Gao X, Song W, Peng X (2015) National innovation demonstration zones leading China's regional development. Mod Econ 6(10):1056
17. Guo R, Guo R, Sangster. (2009) How the Chinese economy works. Palgrave Macmillan, Singapore

18. Han H, Guo L, Zhang J, Zhang K, Cui N (2021) Spatiotemporal analysis of the coordination of economic development, resource utilization, and environmental quality in the Beijing-Tianjin-Hebei urban agglomeration. Ecol Ind 127:107724
19. Harrell S (2007) Recent Chinese history in ecosystem perspective. In: Paradigms influx: new perspectives on shifting grounds in contemporary China and Chinese studies conference. University of California, San Diego, pp 20–22
20. Hu J, Ouyang T, Wei WX, Cai J (2020) How do manufacturing enterprises construct e-commerce platforms for sustainable development? a case study of resource orchestration. Sustainability 12(16):6640
21. Huang L, Zhang Y, Guo Y, Zhu D, Porter AL (2014) Four dimensional Science and Technology planning: a new approach based on bibliometrics and technology roadmapping. Technol Forecast Soc Chang 81:39–48
22. Huang W, Xi M, Lu S, Taghizadeh-Hesary F (2021) Rise and fall of the grand canal in the ancient Kaifeng city of China: role of the grand canal and water supply in urban and regional development. Water 13(14):1932
23. Hymowitz T (1970) On the domestication of the soybean. Econ Bot 24(4):408–421
24. Jing A (2015) A history of the Great Wall of China. World Scientific, Singapore
25. Jing Y, Gao H (2022) Development countermeasures of Tianjin's modern logistics industry under the background of double cycle strategy. The international conference on artificial intelligence and logistics engineering. Springer, Cham, pp 605–615
26. Le Z, Ning Z (2015) The development and research of china industrial cluster based on supply chain: a case of Beijing-Tianjin-Hebei tourism destination. Open Cybern Syst J 9(1)
27. Lee HF, Jia X, Ji B (2022) Population, wars, and the grand canal in Chinese history. Sustainability 14(12):7006
28. Li C, Meng Y, Li Y, Ge J, Zhao C (2019) Inter-metropolitan land-price characteristics and patterns in the Beijing-Tianjin-Hebei urban agglomeration in China. Sustainability 11(17):4726
29. Li F, Yao N, Liu D, Liu W, Sun Y, Cheng W, Li X, Wang X, Zhao Y (2021a) Explore the recreational service of large urban parks and its influential factors in city clusters–experiments from 11 cities in the Beijing-Tianjin-Hebei region. J Clean Prod 314:128261
30. Li J, Zhang P, He J, Sun D (2021b) The spatial agglomeration and industrial network of strategic emerging industries and their impact on urban growth in mainland China. Complexity
31. Li J, Wan S, Li G (2021c) Analysis of the role of government in the construction of cross-regional innovation system. In: E3S Web of conferences, vol. 292. EDP Sciences
32. Li M, Mao C (2020) Spatial effect of industrial energy consumption structure and transportation on haze pollution in Beijing-Tianjin-Hebei region. Int J Environ Res Public Health 17(15):5610
33. Li Z, Zhu C, Wu G, Zheng C, Zhang P (2015) Spatial pattern and temporal trend of prehistoric human sites and its driving factors in Henan Province, Central China. J Geog Sci 25(9):1109–1121
34. Lin JY (2011) New structural economics: a framework for rethinking development. World Bank Res Obs 26(2):193–221
35. Liu G, Liang C, Gong X (2022) Coupling relationship between financial inclusion and agricultural green total factor productivity: mechanism analysis and empirical research. Available from: https://assets.researchsquare.com/files/rs-1619063/v1_covered.pdf?c=1654619070
36. Liu L, Chen X (2012) The archaeology of China: from the late Paleolithic to the early Bronze Age. Cambridge University Press, Cambridge
37. Liu X, Hunt HV, Jones MK (2009) River valleys and foothills: changing archaeological perceptions of North China's earliest farms. Antiquity 83(319):82–95
38. Lu Y, Cao K (2019) Spatial analysis of big data industrial agglomeration and development in China. Sustainability 11(6):1783
39. Lu W, Tam VW, Du L, Chen H (2021) Impact of industrial agglomeration on haze pollution: new evidence from Bohai Sea economic Region in China. J Clean Prod 280:124414

40. Mao X, Jia J, Liu C, Hou Z (2005) A simulation and prediction of agricultural irrigation on groundwater in well irrigation area of the piedmont of Mt. Taihang, North China. Hydrol Process: Int J 19(10):2071–2084
41. Okazaki K, Wei D, Zhu H (2013) Variations in the oral health of millet agriculturalists in the northern 'Great Wall'region of China from the Middle Neolithic to the Sixteen Kingdoms period. Anthropol Sci 130929
42. Ouyang X, Jian Q, Jiang Z (2022) Energy efficiency of the industrial sectors in Beijing-Tianjin-Hebei urban agglomeration: does technological gap matter? Environ Sci Pollut Res 1–14
43. Pearson R, Underhill A (1987) The Chinese Neolithic: recent trends in research. Am Anthropol 89(4):807–822
44. Rhodes M (ed) (1995) The regions and the new Europe: patterns in core and periphery development (Vol. 1). Manchester University Press, Manchester
45. Shan Y (2016) The pattern of archaeological cultures in northern China during the Eastern Zhou Period to the Qin Dynasty–also on the interactions among the Rong, Di and Hu ethnic groups and the Central Plains. Chin Archaeol 16(1):178–188
46. Song G, Childs-Johnson E (2020) Early and middle shang periods. The Oxford Handbook of Early China, p 253
47. Tang G, Wang Y, Li X, Ji D, Hsu S, Gao X (2012) Spatial-temporal variations in surface ozone in Northern China as observed during 2009–2010 and possible implications for future air quality control strategies. Atmos Chem Phys 12(5):2757–2776
48. Tang Z, Zhang Z, Zuo L, Wang X, Hu S, Zhu Z (2020) Spatial econometric analysis of the relationship between urban land and regional economic development in the Beijing–Tianjin–Hebei coordinated development region. Sustainability 12(20):8451
49. Wang Z, Liang L, Sun Z, Wang X (2019) Spatiotemporal differentiation and the factors influencing urbanization and ecological environment synergistic effects within the Beijing-Tianjin-Hebei urban agglomeration. J Environ Manag 243:227–239
50. Wu SH (2012) Alliance-building at the dawn of chinese civilization. J MilY Strat Stud 14(3 & 4)
51. Wu W, Zhang M, Ding Y (2020) Exploring the effect of economic and environment factors on PM2. 5 concentration: a case study of the Beijing-Tianjin-Hebei region. J Environ Manag 268:110703
52. Xing H (2020) Strategies for developing the Beijing-Tianjin-Hebei region to achieve complementary advantages and high-quality development. China Economic Transition = Dangdai Zhongguo Jingji Zhuanxing Yanjiu 3(4):104–114
53. Xu C, Dong L, Yu C, Zhang Y, Cheng B (2020) Can forest city construction affect urban air quality? The evidence from the Beijing-Tianjin-Hebei urban agglomeration of China. J Clean Prod 264:121607
54. Yang P, Mao R, Shao H, Gao Y (2009) The spatial variability of heavy metal distribution in the suburban farmland of Taihang Piedmont Plain, China. Comptes Rendus Biol 332(6):558–566
55. Zhang J, Zhu W, Zhao F, Zhu L, Li M, Zhu M, Zhang X (2018) Spatial variations of terrain and their impacts on landscape patterns in the transition zone from mountains to plains—a case study of Qihe River Basin in the Taihang mountains. Sci China Earth Sci 61(4):450–461
56. Zhang N, Qi X (2020) Development Strategies of Tianjin's Cultural and Creative Industry (CCI) Based on the Beijing-Tianjin-Hebei coordinated development. In: International conference on intelligent and interactive systems and applications. Springer, Cham, pp 393–404
57. Zhang Q (2015) An introduction to Chinese history and culture. Springer, Heidelberg
58. Zhang X, Zhao T, Wang J, Wei Y (2021) The embodied CO_2 transfer across sectors of cities in Jing-Jin-Ji region: combining multi-regional input–output analysis with complex network analysis. Environ Sci Pollut Res 28(32):44249–44263
59. Zhao S, Zhao K, Yan Y, Zhu K, Guan C (2021) Spatio-temporal evolution characteristics and influencing factors of urban service-industry land in China. Land 11(1):13
60. Zhao Z, Lv Z (2009) Global supply chain and the Chinese auto industry. Chin Econ 42(6):27–44
61. Zhou L, Tian L, Cao Y, Yang L (2021) Industrial land supply at different technological intensities and its contribution to economic growth in China: a case study of the Beijing-Tianjin-Hebei region. Land Use Policy 101:105087

62. Zhou Z, Zhang A (2021) High-speed rail and industrial developments: Evidence from house prices and city-level GDP in China. Transp Res Part A: Policy Pract 149:98–113
63. Zhu D, Xu B (2021) Regional government R&D investment and innovation performance: the moderating effect of geographical and organizational proximities. Int J Innov Sci

Chapter 4
Yangtze River Delta City Cluster: From Integrated Production and Services to Digital Technology Innovation

4.1 A Brief Introduction to the Yangtze River Delta City Cluster

The Yangtze River Delta City cluster (hereinafter referred to as the "YRD") is located in the alluvial plain before the Yangtze River into the sea [91]. According to the "Yangtze River Delta Urban Agglomeration Development Plan" approved by the State Council of China, the city cluster in the region includes: Shanghai; Nanjing, Wuxi, Changzhou, Suzhou, Nantong, Yancheng, Yangzhou, Zhenjiang, and Taizhou in Jiangsu Province; Hangzhou, Ningbo, Jiaxing, Huzhou, Shaoxing, Jinhua, Zhoushan, and Taizhou in Zhejiang Province; and Hefei, Wuhu, Maanshan, Tongling, Anqing, Chizhou, and Xuancheng in Anhui Province [21]. The Yangtze River Delta urban agglomeration covers an area of 211.7 thousand square kilometers, accounting for only 2.2% of China's total area. With a population of 225 million, the Yangtze River Delta urban agglomeration has a total GDP of 20.5 trillion RMB in 2020, accounting for 20.18% of China's GDP. It ranks first among all city clusters with a per capita GDP of up to 131,900 RMB [20, 22, 77], much higher than other urban agglomerations in China. Among them, Shanghai's GDP ranks first in the mainland cities (i.e., 159,400 RMB in 2019). Suzhou, Hangzhou, Nanjing, Ningbo, Wuxi, Hefei, and Nantong are all over one trillion RMB.

4.1.1 The Cluster's Significance: A World-Class Urban Agglomeration and China's Window to the World

The Yangtze River Delta (YRD) urban agglomeration is an important intersection of the Belt and Road Initiative and the Changjiang Economic Belt and plays a pivotal strategic role in China's overall modernisation drive and opening up [68, 90]. It is an important platform for China to participate in international competition, an important engine for economic and social development, a leader of the Yangtze

A. Cheshmehzangi and T. Tang, *China's City Cluster Development in the Race to Carbon Neutrality*, Urban Sustainability, https://doi.org/10.1007/978-981-19-7673-5_4

River Economic Belt, and one of the regions with the best urbanisation foundation in China [84]. The economic hinterland of the YRD city cluster is vast, with modern river-sea ports and airports, a relatively sound highway network, and a national leader in the density of highway and railway trunk lines [74, 92, 101], and a comprehensive three-dimensional traffic network [26, 37, 103] has been basically formed.

The Yangtze River Delta Urban Agglomeration Development Plan points out that the urban agglomeration of the Changjiang Delta should build a world-class urban agglomeration facing the world, radiating the Asia–Pacific region and leading the country [22]. The plan aims to build a resource allocation center with the most economic vitality, a scientific and technological innovation highland with global influence, and a crucial modern service and advanced manufacturing center in the world. It is also planned to be recognised as a key international gateway in the Asia–Pacific region, a pioneer in the new round of reform and opening up across the country, and a demonstration zone for constructing a beautiful China [21]. In terms of cities, the permanent population of the three provinces and one city in the Yangtze River Delta is mainly concentrated in Shanghai, Suzhou, Hangzhou, Wenzhou, Xuzhou, Ningbo, Nanjing, Fuyang, and Hefei, which account for 43% of the resident population.

Shanghai had the largest permanent population as per the earlier records in 2019. The number of permanent residents in 2017 was negative, and the city's permanent residents have increased for two consecutive years [65]. Suzhou and Hangzhou have more than 10 million permanent residents as two main megacities in the region (apart from Shanghai as a super megacity), with 10.75 million and 10.36 million, respectively. The permanent population of Wenzhou, Xuzhou, Ningbo, Nanjing, Fuyang, and Hefei exceeds 8 million. Nantong, Yancheng, Wuxi, Taizhou, Suzhou, Jinhua, Bozhou, and Shaoxing have a resident population of between 5 and 8 million. Tongling, Chizhou, Huangshan, and Zhoushan have less than 2 million residents. Therefore, we see a very good mix of cities of various triers and sizes. Compared with 2020, Hangzhou's resident population increment reached 554,000, followed by Ningbo, with 340,000 [65]. Hefei's resident population was 8.189 million, an increase of 102,000 over the years 2020 and 2021 [65]. In addition, the permanent population increment of Jiaxing, Nanjing, Fuyang, and Wenzhou is over 50,000, with 74,000, 63,800, 52,000, and 50,000, respectively [64]. Therefore, as the data shows, there is a growing population flow into the YRD city cluster. In recent years, the YRD city cluster has attracted many young professionals, particularly those seeking new professional job opportunities immediately after graduation. The region and (particularly) its larger cities have become a major hub for job opportunities for the youngsters. It has become a magnet point in the county, where a world-class lifestyle and living are expected.

4.2 A Historical Analysis of the Yangtze River Delta

The Yangtze River Delta region has a long history of culture, coupled with a developed water system, rich land better than most of China's agriculture [45], and handicrafts, so that in the middle and late feudal society (in China) has initially formed a considerable city group [21, 39]. From the Ming Dynasty to the Qing Dynasty, the Yangtze River Delta had nine large commercial and handicraft cities with the textile industry and its trading centers, Nanjing, Hangzhou, Suzhou, and Songjiang [25, 34, 41, 74, 78]. Other major industries were grain distribution centers in Yangzhou, Wuxi, and Changzhou,and cities like Huzhou for printing and stationery production trading centers. At the beginning of the Yuan Dynasty, when counties were just set up, Shanghai became an important commercial hub [30, 32, 85, 94]. At that time, Shanghai developed into an important commercial center of coastal North–South trade.

After the Opium Wars, and specifically from 1842 to 1949, under the condition of opening to the outside world, the YRC became the initial period of great development of the product economy [61, 73, 100]. At that time, the YRD region emerged as a modern industrial and commercial city group formation and development stage [81]. Then, foreign goods began to pour into China. In return, Chinese raw materials began to export, and foreign trade and financial institutions began to enter China [99]. At the same time, the early modern industry of import substitution began to develop. In the second half of the nineteenth century, Shanghai had become an industrial and commercial metropolis through large-scale infrastructure construction [31, 94, 97]. By the 1930s, Shanghai had become the trade center, financial center, and industrial center of the whole Yangtze River Delta and even China [48, 83, 99]. At the same time, Wuxi, Nantong, and Ningbo, also through the superior geographical location of foreign trade to Hong Kong and Xingcheng, became important secondary cities in the region [12, 59]. Meanwhile, cities like Hangzhou, Zhenjiang, Yangzhou, Suzhou, Changzhou, and other older cities [46, 93] were affected due to the decline of the Grand Canal.

In the era of the planned economy (1949–1978), during the period of convergence of urban functions in the Yangtze River Delta, China chose the Soviet-style highly centralized planned economy system and closed economic development strategy under various special circumstances [28, 36, 62, 105]. Every city runs industry uniformly and transforms a consumption city into a production city [56], which makes the city's function converge and the urbanization process slow.

During the period of reform and opening up and the development of a socialist market economy, the urban functions of a city cluster in the Yangtze River Delta region were re-differentiated and reorganized [71]. At that time, Shanghai, with its superior geographical location, became an international metropolis again. Due to transitions over a few decades, the economic center, the tertiary industry relationship, and the urban function of the whole urban agglomeration have been repositioned and divided [22]. The development of key cities in the YRD city cluster is based on transportation and other infrastructure development [9, 57, 98, 104]. The region heavily relies on ports, waterways, transportation hubs, and other important locations [44, 99]. Also, at the same, Shanghai became a center of trade, finance, and information development.

In May 2010, the State Council officially approved the implementation of the "Yangtze River Delta Regional Plan" [49, 88]. The scope of the Yangtze/Changjiang Delta is defined as Jiangsu, Zhejiang, and Shanghai. This is marked by the precise strategic positioning of the region's development. This strategic position is more related to international trade and regional cooperation. For instance, the Asia–Pacific region is an important international gateway [40], known as the world's important modern service industry and advanced manufacturing center. Therefore the cooperation between the YRD and the Asia–Pacific region is crucial. Thus, the YRD city cluster plays a strong role in the international competitiveness of a world-class city group. The 2010 Report on the Development of China's City cluster suggests that China will build the Yangtze River Delta into a world-class city group [64] with the most robust and comprehensive competitiveness in the country after 10–20 years of construction.

In addition, the migrated population to the YRD region helped to promote economic and social development. This development cannot occur without the role of talent acquisition. In recent years, three provinces and one city in the YRD have issued a "golden talent policy", attracting a large number of people [11, 88]. In 2019, the total permanent population of the three provinces and one city in the Yangtze River Delta were approximately 226.2034 million. They had an increase of 1.7896 million compared to the previous year. Among them, Jiangsu's resident population had increased by 193,000 people compared to last year. Anhui resident population of 63.659 million had an increase of 423,000 from a year earlier. Zhejiang's resident population of 58.5 million increased by 1.13 million compared to 2018. Shanghai's resident population was 24,281,400 and experienced an increase of 43,600 compared to the year before [2]. All these recent figures indicate how high-tech and high-level talents are attracted side-by-side with the influx of labour populations, international companies, and young professionals.

4.3 Industrial Structure in the Yangtze River Delta

In this section, we explore the present situation of the industrial cluster in the YRD urban agglomeration. The industrial cluster refers to the small and medium-sized enterprises (SMEs) that are geographically close to, in the same place, or related to a specific industry field, gathering together in high density. In China, it is also called the "industrial cluster," "characteristic industrial zone", and "massive economy" [79, 80]. From theory and practice, the industrial cluster is a vital carrier to promoting the competitiveness and innovation ability of regions and industries. Most industrial clusters in China are mainly small and medium-sized private enterprises [22]. Therefore, it is of great practical significance to promote industrial clusters' sustainable and healthy development [18].

In the Yangtze River Delta region, industrial clusters are distributed in Zhejiang, Jiangsu, Shanghai, and other provinces and cities. Among them, the development of industrial clusters in Zhejiang Province is dominated by small and medium-sized private enterprises [1, 69]. In this area, industries are mainly concentrated in textile, hardware, clothing, and other traditional industries with a certain history [35, 67, 76]. This is known as a typical model of spontaneous growth for industrial clusters. The geographical conditions, business tradition, and the spirit of the Zhejiang people (commonly referred to as the "Zhejiang Spirit") [3] are the important conditions for industrial clusters [50, 82]. Moreover, the development of the traditional local economy and the gathering of mobile workers [95, 109] provide a solid foundation for cluster development. There are some representative models of industrial cluster development in Jiangsu Province [22]. One is the model of southern Jiangsu represented by Suzhou, Wuxi, and Changzhou. It is a spontaneous growth model dominated by collective and medium-sized enterprises. Second, through establishing Suzhou Singapore Industrial Park, Suzhou new area, Wuxi high-tech development zone, Changzhou high-tech development zones, and other state-level development zones, we see the emergence of a planning-oriented industrial cluster model [22]. In this regard, Jiangsu Province has formed leading enterprise-driven industrial clusters. In return, the evidence shows that recent and ongoing projects contribute to regional industrial restructuring. Consequently, industrial parks support the development of local and regional industrial patterns.

Regarding industry distribution, the industrial clusters of Zhejiang and Jiangsu Provinces are mainly concentrated in the traditional dominant and planned high-tech industries. These industries are mostly textile, clothing, machinery, light industry, metallurgy, equipment, and other industries [41]. At present, more than 50 textile and garment clusters have been formed in Jiangsu Province [22]. The clusters are characterised by prominent specialisation, complete industrial chain system, and remarkable agglomeration effect of the SMEs. Lastly, reports show that Shanghai is an important economic center in China, with multinational enterprises gathering and a strong research and development force [22]. It already has petrochemical, fine steel, automobile manufacturing, electronic information, biomedicine, shipbuilding and equipment, and a number of industrial clusters of varying sizes [64]. In terms of industry distribution, industrial clusters in Shanghai are mainly concentrated in the main pillar industries with competitive advantages in the country, reflecting the characteristics of large industries and large clusters [22, 52, 102]. In sum, the industrial structure of the YRD is complicated yet includes prominent industries linked to national and international level cooperation, trade, and business.

4.3.1 Industrial Structure Optimisation

Since the reform and opening-up period, the YRD has vigorously developed export-oriented industries, which has effectively promoted regional economic and social development [22]. This effect also facilitated the rise of cities in the region. At present,

China-US economic relations are in an unstable state. The COVID-19 pandemic has not yet been effectively controlled. Other barriers also exist that show growing tension and decline for larger economies. More importantly, world economic development is at a crossroads [108]. Thus, the future industrial structure of the Yangtze River Delta region should be based on regional and domestic demand [53], accelerate the application of new generation of information technology or IT/ICT, and promote the integration of traditional industries and digital technology [22]. On the basis of the emerging dual cycle development pattern, domestic circulation has become the main factor for international and domestic mutual promotion [75, 96]. In accordance with the requirements of transformation and upgrading of regional demand structure, the YRD accelerates the construction of industrial clusters based on regional and domestic demand [54, 112]. It also promotes the integrated development of manufacturing and service industries and constantly optimise the regional industrial structure [15, 23, 111]. More recently, the plan has shifted to develop the industrial foundation for the Yangtze River Delta region and toward building a global city region. Hence, the YRD has been in the global economic competitiveness, representing China's economic power on the global stage.

As mentioned earlier, much of the industrial structure optimisation in the YRD city cluster is based on regional and domestic demand. In particular, the COVID-19 pandemic has boosted local demand for most industries in China. Since the reform and opening up period, under the guidance of China's opening policy, the YRD has vigorously developed an export-oriented economy [107], which has effectively promoted regional economic and social development [17, 42, 70, 79, 80]. Since then, foreign trade has become an important engine of economic growth in the Yangtze River Delta [22]. Many provinces and cities have reached a high level of dependence on foreign trade. The YRD region has participated in the international economic cycle, with export demand as an important driving force of economic growth. This approach stimulates regional investment demand and consumer demand, forming a comprehensive force to promote economic growth and promote the rise of the Changjiang Delta urban agglomeration [22]. However the export-oriented economic growth model of the Yangtze River Delta region also influenced different cities in the region to form a "single fight" and "multiple external" production pattern [13, 27, 51]. Foreign demand-oriented production leads to regional production supply and consumer/consumption demand, which is uually inconsistent. Under the pattern of "multi-headed external" production, many factors are explored [10, 16, 60, 106]. For instance, the problems of division of labour and cooperation, competition and cooperation between cities, corresponding problems of production and consumption, and perfect and defective industrial chains [29, 33]. Moreover, industrial isomorphism has been prominent in the YRD, showing the need to optimise the industrial structure. These problems are puzzling the economic integration of the YRD [22] and also hindering the region from participating in the international economic competition with the overall advantage.

At this stage, economic globalisation is greatly restricted by the current epidemic and international trade situation. Since 2019, such a situation has directly affected the growth of imports and exports in the YRD region. On the other hand, the global international trade crises will promote the adjustment of the industrial structure in the YRD [22]. At present, a new pattern of double-cycle development is taking shape, which is dominated by domestic circulation and promoted by international and domestic cooperation [58, 75]. In addition to medical supplies, exports of the YRD region to stimulate economic growth will weaken in the next period. In the global crisis, the Yangtze River Delta region (i.e., according to the actual situation to adjust and optimise the industrial structure) could solve some of the existing problems [22]. Hence, the regional stakeholders believe the future can get better development prospects.

Lastly, we argue that the YRD economic production guidance should be timely adjusted. In this regard, economic production should focus on regional and domestic consumption and production demand. By focusing on domestic and regional demand, the YRD continues to improve the matching of production and consumption [22]. It will also help enhance the coupling between supply and demand and promote enterprises to strengthen the horizontal and vertical relationship between each other. According to the plans, the primary goals are to build an innovation chain, talent chain, the information chain, and capital chain between enterprises [21]. In return, the plan could improve the industrial chain and form industrial clusters. Based on the YRD region and China's domestic consumer demand, the development of regional industries is the backbone of the next steps. For instance, the formation of a regional industrial system could build a more reasonable regional industrial structure, which is essential for the next steps of carbon intensity, carbon peak, and carbon neutrality plans. As described earlier, enterprises and cities should pay attention to the formation of the division of labour and cooperation as well as competition and cooperation with surrounding enterprises and surrounding cities [63]. They should reduce the degree of industrial isomorphism and get rid of the city "alone" and "many external" modes of economic competition [43]. Such an approach should improve the level of cooperation between cities and promote the collaborative innovation of science and technology in the YRD region. The following steps ought to continuously strengthen the regional and domestic economic cycle. In doing so, the enterprises and cities can promote economic integration in the region and then participate in international economic competition with the overall advantages of regional and urban groups [22]. In this way, the YRD region can participate in the international economic cycle with a better outlook and higher competitiveness. It also can play a more effective role at the international level and participate in the next round of the economic globalisation process.

4.4 Sustainable Development Plan: Development of Integration of Production and Services

Since the 1990s, the YRD region has stepped into a new development phase. For Shanghai, for instance, the Pudong development was a remarkable new era [71]. Also, the opening effect driven by economic and social development into a fast development track helped the region to enter a new development era. At present, the economy of the YRD region is changing from the stage of high-speed growth to the phase of high-quality development. It is in the critical step of changing the development mode and optimising the industrial structure. The YRD region's per capita GDP and disposable income have reached the leading level in China [22]. People's demand structure is in the stage of transformation and upgrading, and people's growing needs for a better life are constantly enhanced. At the same time, the ongoing epidemic (i.e., since late 2019) has made people more concerned about the health of individuals and family members. Thus, more attention is given to public health and hygiene issues. In short, the problem of upgrading consumer demand is becoming the theme of industrial development in the YRD region.

Economic production in the region pays more attention to personal development and needs, and treats different needs of different groups differently. This economic production model is expected to be the development trend of consumption demand in the YRD. Regional industries that can adapt to consumption upgrades can be further developed, feeding into industrial clusters and chain development. With the upgrading of consumer demand, mass production of material goods and concentrated service activities will give way to small-scale, family-oriented, and individual-oriented economic production activities [22]. Small batch production, personalised services, and products that meet advanced needs will be developed [21, 22]. For further popularities of industries in the region, the focus will be on the material contents of products and provided external services. In the coming decade or so, and by the team, China is expected to reach its carbon peak, and production and services of the YRD are expected to be in the form of integrated development. The production of material products in the primary and secondary industries is still very important [22], including food and industrial products. At the same time, the material products produced by the agriculture and food industry need to be adequately packaged and served. Material production must contain service content, which is expected to boost service quality in the region and for the optimisation of industrial structure. Suppose companies do not attach importance to the after-sales service of productive enterprises. In that case, they are expected to struggle to meet the demand and pathways of development at the regional level [22]. As proposed in the governmental plans, the boundary between material products and service products will be increasingly blurred [64]. In the future, many products will be confused as agricultural, industrial, or service products according to the current classification. Many industries will develop from industrial separation to industrial integration. This approach will benefit the YRD from economic and environmental perspectives. Also, if we look at it institutionally [35], then we see the ongoing restructuring could benefit regional carbon neutrality targets, too.

Furthermore, the upgrading of consumption demand is reflected in the production link, which is the integration of material production and production services [21]. Material production presents the trend of service industrialisation [2], which is key to the production and provision of services. In the process of production, the proportion of intermediate input of services in the production process is higher, which is manifested in R&D, design, finance, leasing, business services, and logistics [22]. On the other hand, service enterprises make use of their own technology, management, and channel advantages, through Original Equipment Manufacturers (OEMs) and chain operation in the material production enterprises, in order to make products more in line with consumer demand [6, 55]. The new format and mode formed through the deep integration of production and services will be able to better meet people's growing needs for a better life, better cope with the problem of insufficient foreign demand, and promote optimisation [89] and upgrading of regional industrial structure and high-quality economic development. These factors are all in line with the regional-level sustainable development strategies and plans, which are mostly economically driven for now.

4.4.1 Optimising the Deep Application of Digital Technology

Digital technology is the fourth industrial revolution after the revolution of the steam engine, electricity, and information [4]. With the aid of digital technology, the level of industrial technology and industrial structure in the Yangtze River Delta can be greatly improved. At present, the development of the digital industry in YRD is close to that of developed countries. The digital industry based on the Internet is becoming more critical to the industrial development of YRD. During the COVID-19 pandemic, the YRD region boosted many digitalisation and digitisation strategies that were tested through integrated models and standalone methods to optimise digital technology's deep application. Such approaches innovated some extraordinary development in digital industries, such as online offices, online education, and e-commerce [47, 110]. There are also some industries through information technology and intelligent technology transformation to boost rapid development, such as online supermarkets, online games, online cinema/entertainment, telemedicine, and so on. Some traditional supermarkets, pharmacies, and medical treatment with the help of internet technology transformation already boost their online sales and service platforms. These approaches are beyond the space constraints and have been the key drivers of rapid development in the YRD region. Some traditional industries, such as agriculture and manufacturing, accelerate the integration of online display and offline production [22]. They actively use the new online economy, display their products, and improve product sales. Some enterprises use online and remote communication to reduce unnecessary travel and regular working hours. In the YRD region, information technology in the sales channel also plays a major role. All these transformations boosted more recently due to the COVID-19 pandemic [14–16] have helped the YRD to progress in adaption and innovation in digital technology integration.

However, from the overall point of view, the digital industry and traditional industries in the Yangtze River Delta region could be described as the "two skin" phenomenon [24], meaning that the degree of integration of the two needs to be further improved. The digital integration industry in the YRD region is still relatively low in GDP [22]. Traditional industries, such as agriculture, animal husbandry, manufacturing, construction, and services, still account for a high proportion of the YRD's economy. Their informatisation, automation, standardisation, and sharing level is very low, far below the level of digitisation and automation in industries such as Germany [8, 22, 38]. In the YRD region, in between agriculture and the manufacturing industry, different technology and industrial forms coexist with each other. They have not carried out a more thorough digital upgrading and information technology revolution [66]. In many traditional industries, enterprise investment in digital technology awareness is relatively weak. Therefore, enterprise industrial software development is relatively slow, and digital industry transformation is far from complete [22]. Also, as a critical matter, the industrial Internet in the YRD region is still at the preliminary (or perhaps primary) stage of development. Consequently, the network system of the scientific and technological service platform is still under construction.

At present, the traditional industries in the YRD, such as medicine, chemicals, machinery, machine tools, automobiles, etc., still occupy an important position in the regional economic output value [22]. The material production industry still plays a vital role in the economic and social development of the Yangtze River Delta [22]. The promotion and integration of traditional and digital industries in the YRD are of great significance for improving industrial technology and optimizing industrial structure. It is also a problem that must be solved for the further development of the Yangtze/Changjiang Delta. Furthermore, the YRD industries need to use modern artificial intelligence technology and the digital economy to digitally transform and upgrade the whole process of new product design, development, mass production, and sales. The same industry can share some technical data, saving research and development costs. At this stage, the YRD region is accelerating the pace of "new infrastructure", to strengthen the big data, artificial intelligence (AI), industrial internet upgrading of traditional industries, and improve quality and efficiency [7, 104]. These are ongoing processes to promote the industrial technology level in the region, which would also help improve the industrial structure and the overall restructuring.

4.5 Pathways to Carbon Neutrality in the Yangtze River Delta City Cluster

In short, based on the current domestic and international economic situation, the industrial development of the YRD region should be based on regional and domestic demand. In response to the demand for upgrading regional consumption, the local governments actively promote the integration of production and services, further develop and meet regional consumption demand, accelerate the transformation and upgrading of traditional industries with digital technology, and promote the optimisation of regional industrial structure. In recent years, plans and strategies for carbon neutrality targets have become key local-level and regional-level tools of change. We see at the local and regional levels no longer experimental projects [17], but fast-developing and large-scale projects (Cheshmehzangi et al. 2018) that are leading beyond just low carbon development plans.

A few large-scale initiatives and programmes would suit the YRD's approach to achieving the carbon–neutral targets. For instance, the ongoing restructuring has a high potential for carbon sinks and emission reduction. The carbon neutral potential of forests is an example that focuses on climate goals [72]. High-quality economic development is also provided as a significant pathway [87], which includes both quantitative and qualitative aspects. In addition, the spatiotemporal aspect of the region plays a part in how land use policy is developed for the carbon peak targets of 2030 and the carbon neutral goals of 2060 [5, 86]. In sum, we anticipate more attention to energy conservation strategies, urban renovation plans, a high-quality urbanisation approach, and digital technology integration. In larger-scale projects like logistics and smart transportation, we see tremendous potential for low-carbon (and eventually zero-carbon) consumption. On the other hand, for production, the imposed strategies may need to expand beyond just the current industrial restructuring and towards decarbonisation and rapid emission reductions. Lastly, we note that due to the global importance of the YRD city cluster, the industrial and economic aspects of the region could play a dual role in how the carbon neutrality targets are met or achieved.

References

1. Akoorie ME, Ding Q (2009) Global competitiveness in the Datang hosiery cluster, Zhejiang. Chin Manag Stud. Available from: https://www.emerald.com/insight/content/doi/https://doi.org/10.1108/17506140910963611
2. Baike (2021) Yangtze River Delta urban agglomeration. 长江三角洲城市群. Available from: https://baike.so.com/doc/5613951-5826561.html
3. Bellandi M, Lombardi S (2012) Specialized markets and Chinese industrial clusters: the experience of Zhejiang Province. China Econ Rev 23(3):626–638
4. Berger R (2014) Digital technology is the fourth industrial revolution, Multimedia Medien, article by Roland Berge

5. Cai W, Peng W (2021) Exploring spatiotemporal variation of carbon storage driven by land use policy in the Yangtze River Delta Region. Land 10(11):1120
6. Candace YY, Ngai EWT, Moon KL (2011) Supply chain flexibility in an uncertain environment: exploratory findings from five case studies. Supply Chain Manag: Int J
7. Chan S (2021) Inter-regional development in China: an assessment. J Infrastruct, Policy Dev 5(1):1270
8. Chang V, Xu YK, Zhang J, Xu Q (2020) Research on intelligent manufacturing development approach for China's local valve industry. Smart Sustain Built Environ
9. Chen H, Zhang Y, Zhang N, Zhou M, Ding H (2022) Analysis on the spatial effect of infrastructure development on the real estate price in the Yangtze River Delta. Sustainability 14(13):7569
10. Chen JY, Qiu JL (2019) Digital utility: Datafication, regulation, labor, and DiDi's platformization of urban transport in China. Chin J Commun 12(3):274–289
11. Chen W, Huang X, Liu Y, Luan X, Song Y (2019) The impact of high-tech industry agglomeration on green economy efficiency—evidence from the Yangtze river economic belt. Sustainability 11(19):5189
12. Chen Z (2011) Modern Chinaâ s network revolution: chambers of commerce and sociopolitical change in the early twentieth century. Stanford University Press, Redwood City
13. Cheng JYS, Ngok KL, Huang Y (2012) Multinational corporations, global civil society and Chinese labour: workers' solidarity in China in the era of globalization. Econ Ind Democr 33(3):379–401
14. Cheshmehzangi A (2020) The city in need: urban resilience and city management during disruptive disease outbreaks. Springer, Singapore
15. Cheshmehzangi A (2021a) From transitions to transformation: a brief review of the potential impacts of COVID-19 on boosting digitization, digitalization, and systems thinking in the built environment. J Build Constr Plan Res 9(1). https://doi.org/10.4236/jbcpr.2021.91003
16. Cheshmehzangi A (2021) Revisiting the built environment: 10 potential development changes and paradigm shifts due to COVID-19. J Urban Manag 10(2):166–175
17. Cheshmehzangi A, Chen H (2021) Key suggestions and steps ahead for China's carbon neutrality plan. In: China's sustainability transitions. Springer, Singapore, pp 189–206
18. Cheshmehzangi A, Dawodu A, Sharifi A (2021) Sustainable urbanism in China. Routledge, New Yok
19. Cheshmehzangi A, Li H, Yang X (2018) Low carbon and climate resilient urban development in China. ADB Report, developed for the National Development Reform Committee (NDRC), Beijing, China
20. Cheshmehzangi A, Li Y, Li H, Zhang S, Huang X, Chen X, Su Z, Sedrez M, Dawodu A (2022) A hierarchical study for urban statistical indicators on the prevalence of COVID-19 in Chinese city clusters based on multiple linear regression (MLR) and polynomial best subset regression (PBSR) analysis. Sci Rep 12(1):1–16
21. China Business Industry Research Institute (2020a) Focus on Yangtze River Delta integration: 2020 Changjiang Delta, three provinces and one city resident population list (Figure). 聚焦长三角一体化:2020年长三角三省一市常住人口排行榜 (图). Available from: https://www.askci.com/news/data/hongguan/20200824/1048251196314.shtml
22. China Business Industry Research Institute (2020b) Analysis of industrial cluster status and distribution in major provinces of Yangtze River Delta in 2020 (Chart). 2020年长三角地区主要省份产业集群现状及布局分析 (图表). Available from: https://www.askci.com/news/chanye/20200521/1417511160684.shtml
23. Du W, Yang Y (2022) The coordinated development of manufacturing industry and logistics industry in the Yangtze River economic belt: empirical study by stages based on Haken model. PLoS ONE 17(2):e0263565
24. Duan P, Zhang L, Song K, Han X (2020) The Push by 5G. In: Communication of smart media. Springer, Singapore, pp 119–135
25. Fu C, Cao W (2019) Transformation of Chinese cities in the early qing dynasty. In: Introduction to the Urban History of China. Palgrave Macmillan, Singapore, pp 305–347

26. Gao S (2021) Analysis on bulk cargo transportation organization and ship transportation system optimization of Yangtze River. In: Fifth International Conference on Traffic Engineering and Transportation System (ICTETS 2021), vol. 12058. SPIE, pp 277–283
27. Ge Y, Guan L (2015) The transformation from factor input-driven growth model to innovation-driven growth model. In: Transition of the Yangtze River Delta. Springer, Tokyo, pp 119–143
28. Glawe L, Wagner H (2021) The rise of China. In: The economic rise of East Asia. Springer, Cham, pp 143–241
29. Gong Q, Guo G, Li S, Liang X (2021) Examining the coupling coordinated relationship between urban industrial co-agglomeration and intensive land use. Land 10(5):499
30. Goodman B (1995) Native place, city, and nation: regional networks and identities in Shanghai, 1853–1937. University of California Press, Oakland
31. Gu FR, Tang Z (2002) Shanghai: reconnecting to the global economy. In: Global networks, linked cities, pp 273–308
32. Guo R (2013) Shanghai. In: Regional China. Palgrave Macmillan, London, pp 262–275
33. Guo S, Ma H (2022) Can the spatial function division of urbanization promote regional coordinated development? evidence from the Yangtze River economic belt in China. Sustainability 14(12):7115
34. Haiyan D, Hong Z, Liang Z (2021).Study on function transformation and spatial reconstructuring of small towns in China's South of the Yangtze River From 1840 to 1949: Case on Changzhou. In: IOP conference series: materials science and engineering, vol 1203, No 2. IOP Publishing, p 022095
35. He C, Wei YD, Xie X (2008) Globalization, institutional change, and industrial location: economic transition and industrial concentration in China. Reg Stud 42(7):923–945
36. Ho BS (2008) The role of local government in China's economic development. London School of Economics and Political Science (United Kingdom)
37. Hu G (2017) Construction cost of China's three-dimensional transportation. In: The cost of development in China. Springer, Singapore, pp 393–407
38. Huang MQ, Ninić J, Zhang QB (2021) BIM, machine learning and computer vision techniques in underground construction: current status and future perspectives. Tunn Undergr Space Technol 108:103677
39. Huang PC (1990) The peasant family and rural development in the Yangzi Delta, 1350–1988. Stanford University Press, Redwood City
40. Hutton TA (2003) Service industries, globalization, and urban restructuring within the Asia-Pacific: new development trajectories and planning responses. Prog Plan 61(1):1–74
41. Ji XD (2003) Concepts of cost and profit in Chinese agricultural treatises: with special reference to Shengshi Nongshu and Pu Nongshu in the seventeenth century. Account Bus Financ Hist 13(1):69–81
42. Jiang Z, Lin Z, Xia T, Wang H, Zheng H (2018) Viewing the development strategy of professional degree graduate education from the demand of China's industrial development. Int J Econ Finance Manag Sci 6(6):290–296
43. Kong Y, Javed F, Sultan J, Hanif MS, Khan N (2022) EMA implementation and corporate environmental firm performance: a comparison of institutional pressures and environmental uncertainty. Sustainability 14(9):5662
44. Lee DS, Liu H, Pan Y (2009) Study on cross-administration innovation system of the Yangtze River Delta. Int J Econ Financ 1(2):213
45. Li B (1998) Agricultural development in Jiangnan, 1620–1850. Springer, Cham
46. Li J, Zhang H, Sun Z (2020) Spatiotemporal variations of land urbanization and socioeconomic benefits in a typical sample zone: A case study of the Beijing-Hangzhou Grand Canal. Appl Geogr 117:102187
47. Li T, Han D, Ding Y, Shi Z (2020) How does the development of the internet affect green total factor productivity? evidence from China. IEEE Access 8:216477–216490
48. Li Y, Jonas AE (2019) City-regionalism as countervailing geopolitical processes: the evolution and dynamics of Yangtze River Delta region, China. Polit Geogr 73:70–81

49. Li Y, Wu F (2018) Understanding city-regionalism in China: regional cooperation in the Yangtze River Delta. Reg Stud 52(3):313–324
50. Liu R, Weng Q, Mao G, Huang T (2013) Industrial cluster, government agency and entrepreneurial development: a case study of Wenzhou City, Zhejiang Province. Chinese Manag Stud
51. Liu T, Yu Z (2022) The relationship between open technological innovation, intellectual property rights capabilities, network strategy, and AI technology under the Internet of Things. Oper Manag Res 1–16
52. Liu X (2012a) An economic analysis of the specialization and marketization of convention and exhibition industry. Int Bus Manag 5(1):86–92
53. Liu X (2012b) Discussing on opening-up policy of coastal regions based on the idea of economy sustainability. In: Technology for education and learning. Springer, Berlin, pp 311–315
54. Liu X, Wang Z, Sun X, Zhang L, Zhang M (2020) Clarifying the relationship among clean energy consumption, haze pollution and economic growth–based on the empirical analysis of China's Yangtze River Delta Region. Ecol Complex 44:100871
55. Liu X, Yang J, Qu S, Wang L, Shishime T, Bao C (2012) Sustainable production: practices and determinant factors of green supply chain management of Chinese companies. Bus Strateg Environ 21(1):1–16
56. Liu Y, Li J, Yang Y (2018) Strategic adjustment of land use policy under the economic transformation. Land Use Policy 74:5–14
57. Luo X, Shen J (2009) A study on inter-city cooperation in the Yangtze River Delta region, China. Habitat Int 33(1):52–62
58. Lv S, Zhao S, Liu H (2022) Research on the coupling between the double cycle mode and technological innovation systems: empirical evidence from data envelopment analysis and coupled coordination. Systems 10(3):62
59. Ma D (2004) Modern economic growth in the Lower Yangzi in 1911–1937: a quantitative, historical and institutional analysis (p. 6). Discussion paper 2004–06–002, Foundation for Advanced Studies on International Development, Tokyo
60. Ma K, Liu M, Zhao T, Zhou J, Yu Y (2021) PTWA: pre-training with word attention for chinese named entity recognition. In: 2021 International joint conference on neural networks (IJCNN). IEEE, pp 1–8
61. McLaren AE (2002) Chinese cultural revivalism: changing gender constructions in the Yangtze River delta. Gender and power in affluent Asia. Routledge, New York, pp 209–235
62. Naughton BJ (2018) The Chinese economy: adaptation and growth. MIT Press, Cambridge
63. Niu F, Yang X, Wang F (2020) Urban agglomeration formation and its spatiotemporal expansion process in China: from the perspective of industrial evolution. Chin Geogra Sci 30(3):532–543
64. People's Daily (2020) Expert interpretation of the new era Yangtze River Delta industrial structure optimization direction. 专家解读新时期长三角地区产业结构优化方向. Available from: http://district.ce.cn/newarea/roll/202007/28/t20200728_35411036.shtml
65. QQ News (2021) Pearl River Delta, Beijing-Tianjin-Hebei, Yangtze Delta-China's three economic circles 2020 GDP report card. 珠三角、京津冀、长三角-中国三大经济圈2020年GDP成绩单. Available from: https://new.qq.com/rain/a/20210320A0C18B00
66. Shen L, Sun C, Ali M (2021) Influencing factors and paths of upgrading consumer goods industry in Shanghai: a FsQCA approach. Int J Emerg Mark
67. Shi L, Ganne B (2009) Understanding the Zhejiang industrial clusters: questions and re-evaluations. In: Asian Industrial clusters, global competitiveness and new policy initiatives, pp 239–266
68. Shira D, Devonshire-Ellis C, Jones SL, Ku E (eds) (2012) The Yangtze River Delta: business guide to the Shanghai region. Springer Science & Business Media, Chan, Singapore
69. Sun Z, Perry M (2008) The role of trading cities in the development of chinese business cluster. Int Bus Res 1(2):69–81

70. Tao Z, Shuliang Z (2022) Collaborative innovation relationship in Yangtze River Delta of China: subjects collaboration and spatial correlation. Technol Soc 69:101974
71. The State Council Information Office (2021) SCO briefing on supporting high-level reform and opening up of Pudong new area and building Pudong into a pioneer area for socialist modernization. Available from: http://english.scio.gov.cn/pressroom/node_8025136.htm
72. Tian H, Zhu J, Jian Z, Ou Q, He X, Chen X, Li C, Li Q, Liu H, Huang G, Xiao W (2022) The carbon neutral potential of forests in the Yangtze River economic belt of China. Forests 13(5):721
73. Wang C, Ducruet C (2013) Regional resilience and spatial cycles: long-term evolution of the Chinese port system (221bc–2010ad). Tijdschr Econ Soc Geogr 104(5):521–538
74. Wang F, Gao C, Hu W (2020) The influence of water transportation evolution on the economic development of cities along the Beijing–Hangzhou Grand Canal since the late Qing Dynasty. In: Water-related urbanization and locality. Springer, Singapore, pp 27–45
75. Wang F, Wang R, He Z (2022) Exploring the impact of "double cycle" and industrial upgrading on sustainable high-quality economic development: application of spatial and mediation models. Sustainability 14(4):2432
76. Wang J (2008) Development of industrial districts in China and its implications. Available from: http://www.firstdraft.it/wp-content/uploads/2008/03/wang-jici_chinese-clusters.pdf
77. Wang X, Ding S, Cao W, Fan D, Tang B (2020) Research on network patterns and influencing factors of population flow and migration in the Yangtze River Delta urban agglomeration, China. Sustainability 12(17):6803
78. Wang Y, Wu W, Boelens L (2021) City profile: Suzhou, China—the interaction of water and city. Cities 112:103119
79. Wei S, Lv X, Cong H (2015) The spatial distribution of industries: from massive economic to industrial cluster. In: Transition of the Yangtze River Delta. Springer, Tokyo, pp 21–48
80. Wei YD (2015) Zone fever, project fever: development policy, economic transition, and urban expansion in China. Geogr Rev 105(2):156–177
81. Wei YD, Leung CK (2005) Development zones, foreign investment, and global city formation in Shanghai. Growth Chang 36(1):16–40
82. Wei YD, Li W, Wang C (2007) Restructuring industrial districts, scaling up regional development: a study of the Wenzhou model. China Econ Geogr 83(4):421–444
83. Wu F (2000) Place promotion in Shanghai. PRC. Cities 17(5):349–361
84. Wu S, Zhang K (2021) Influence of urbanization and foreign direct investment on carbon emission efficiency: evidence from urban clusters in the Yangtze River economic belt. Sustainability 13(5):2722
85. Wu W (1999) City profile: Shanghai. Cities 16(3):207–216
86. Wu W, Zhang T, Xie X, Huang Z (2021) Regional low carbon development pathways for the Yangtze River Delta region in China. Energy Policy 151:112172
87. Xiao R, Pan L, Xiao H, Xiao H, Zhu Z (2022) Research of intelligent logistics and high-quality economy development for Yangtze River cold chain shipping based on carbon neutrality. J Mar Sci Eng 10(8):1029
88. Xu H, Qiu L, Liu B, Liu B, Wang H, Lin W (2021) Does regional planning policy of Yangtze River Delta improve green technology innovation? evidence from a quasi-natural experiment in China. Environ Sci Pollut Res 28(44):62321–62337
89. Yan Z, Ismail H, Chen L, Zhao X, Wang L (2019) The application of big data analytics in optimizing logistics: a developmental perspective review. J Data, Inf Manag 1(1):33–43
90. Yang Y, Lu H, Liang D, Chen Y, Tian P, Xia J, Wang H, Lei X (2022) Ecological sustainability and its driving factor of urban agglomerations in the Yangtze River economic belt based on three-dimensional ecological footprint analysis. J Clean Prod 330:129802
91. Yao Y (2016) The Yangtze River impacts and early Nanjing Geography. In: Nanjing: historical landscape and its planning from geographical perspective. Springer, Singapore, pp 15–38
92. Yin J, Li A (2011) The strategy for integrating the logistics industry of Yangtze River Delta of China (Master's thesis, Høgskolen i Molde)

93. Yimin HE (2008) The decline of traditional industrial and commercial cities in modern China: exemplified with Suzhou, Hangzhou and Yangzhou. Front Hist China 3(2):263–292
94. Yin J, Yin Z, Zhong H, Xu S, Hu X, Wang J, Wu J (2011) Monitoring urban expansion and land use/land cover changes of Shanghai metropolitan area during the transitional economy (1979–2009) in China. Environ Monit Assess 177(1):609–621
95. Yu G, Zhang H (2011) Notice of Retraction: Speed up the strategy of the Internet of Things industry. In: *2011* International conference on E-business and E-government (ICEE). IEEE, pp 1–4
96. Yu G, Zhou X (2021) The influence and countermeasures of digital economy on cultivating new driving force of high-quality economic development in Henan Province under the background of double circulation. Ann Oper Res 1–22
97. Yue W, Chen Y, Thy PTM, Fan P, Liu Y, Zhang W (2021) Identifying urban vitality in metropolitan areas of developing countries from a comparative perspective: Ho Chi Minh City versus Shanghai. Sustain Cities Soc 65:102609
98. Zhang J, Zhang R, Xu J, Wang J, Shi G (2021) Infrastructure investment and regional economic growth: evidence from Yangtze River economic zone. Land 10(3):320
99. Zhang T (2006) From intercity competition to collaborative planning: the case of the Yangtze River Delta region of China. Urban Affairs Rev 42(1):26–56
100. Zhang X (2015) Globalization and the megaregion: investigating the evolution of the Pearl River Delta in a historical perspective. In: Megaregions. Edward Elgar Publishing, pp 175–199
101. Zhang X, Hu Y, Lin Y (2020) The influence of highway on local economy: evidence from China's Yangtze River Delta region. J Transp Geogr 82:102600
102. Zhang Y, Wei J, Gong L (2012) A co-evolutionary process among industrial cluster policy, leading enterprises behavior and industrial cluster innovative capacity: the industry in Quzhou, Zhejiang. In: 2012 Proceedings of PICMET'12: technology management for emerging technologies. IEEE, pp 1926–1935
103. Zhang Z, Hu B, Qiu H (2022) Comprehensive evaluation of resource and environmental carrying capacity based on SDGs perspective and three-dimensional balance model. Ecol Ind 138:108788
104. Zhao J, Yu Y, Wang X, Kan X (2017) Economic impacts of accessibility gains: case study of the Yangtze River Delta. Habitat Int 66:65–75
105. Zheng D (2013) Economic policies, regional inequality, and growth: evidence from China (Doctoral dissertation, 名古屋大学)
106. Zheng X, Cheng T, Yang L, Xu J, Tang J, Xie K, Huang X, Bao Z, Zheng X, Diao Y, You Y (2019) Genetic diversity and DNA fingerprints of three important aquatic vegetables by EST-SSR markers. Sci Rep 9(1):1–10
107. Zheng Y (2010) China, an emerging power, is exploring its own development model. China Econ, Forthcoming. Available from: https://papers.ssrn.com/sol3/papers.cfm?abstract_id=1532106
108. Zheng Y, Tong SY (2010) China and the global economic crisis, vol 22. World Scientific, Singapore
109. Zhong H, Zhang J, Zhang S, Zheng W (2021) Study on spatial imbalance and determinants of E-commerce development in Zhejiang, China. Math Problems Eng
110. Zhou G (2022) The Impact of COVID-19 pandemic on the digital economy and policy responses. In: COVID-19'S economic impact and countermeasures In China, pp 175–197
111. Zhu B, Zhang T (2021) The impact of cross-region industrial structure optimization on economy, carbon emissions and energy consumption: a case of the Yangtze River Delta. Sci Total Environ 778:146089
112. Zou S, Li M, Chen J, Chen Y (2021) Does the construction of an integrated transport network promote urban innovation? a perspective based on the theory of flow space. PLoS ONE 16(11):e0259974

Chapter 5
Pearl River Delta City Cluster: From Dual-Core Structure Economic Development Strategies to Regional Economic Plans

5.1 A General Introduction to the Pearl River Delta City Cluster

Pearl River Delta City cluster, also known as Zhujiang Delta city cluster, is one of China's three major urban agglomerations. The other two are the Yangtze River Delta and the Beijing-Tianjin-Hebei Bohai Sea City cluster. The Pearl River delta urban agglomeration is the most dynamic and has the highest urbanisation rate among all 19 city clusters in China. The Pearl River Delta city cluster is centered in Guangzhou, Shenzhen, and Hong Kong, including Zhuhai, Huizhou, Dongguan, Zhaoqing, Foshan, Zhongshan, Jiangmen, and Macao. On its own, the cluster includes several many economic powers in China.

The Pearl River Delta (PRD) city cluster is located in the south of China, covering 181,000 square kilometers. In 2019, the population of nine cities in the Pearl River Delta was about 64.49 million. The significant population growth from inland and neighbouring provinces to Guangdong province has been a phenomenon that could be tracked back decades ago. China's focus on southern regions, including the two strongholds of Hong Kong and Macao, has enabled many economic opportunities for the region's development. The emergence of this city cluster under the inevitable leadership of the provincial government sets a mark for promoting the regional economic zones at the larger scales of national and international levels. Due to its broader connections with the international market, it also includes "*carbon footprints of foreign regions*…[that]…*are concentrated in key manufacturing hubs*" [58]. The GDP per capita of the PRD is the highest among all urban agglomerations, making it the most prominent city cluster in China (i.e., from the economic perspective). The high-level GDP is recorded to positively impact the region's industrial sector, particularly the energy efficiency of the industrial sector [39]. The per capita GDP in 2019 was approximately 134,800 yuan, 0.87 million yuan higher than that of the Yangtze Delta and Beijing-Tianjin-Hebei clusters. Hence, we can see that the PRD city cluster has grown into a super economic power.

A. Cheshmehzangi and T. Tang, *China's City Cluster Development in the Race to Carbon Neutrality*, Urban Sustainability, https://doi.org/10.1007/978-981-19-7673-5_5

From the perspective of urbanisation, the urbanization rate of the Pearl River Delta reached 85.3% in 2018, accounting for the highest urbanisation rate in China. Guangzhou and Shenzhen, as the "dual-core" cities of the Pearl River Delta, mainly benefit from the reform and opening up. This dual-core structure accounts for 80% of the Guangdong province's GDP [25]. Guangzhou's GDP rose from the eighth in 1978 to the third in 2015 and was surpassed by Shenzhen [2, 13, 17]. Since 2015, its annual population growth has exceeded 400,000, second only to Shenzhen. Even in the population forecasting studies, like the one by [6], we expect to see a similar trend till at least reaching the carbon peak timeline in 2030. On the one hand, we have the historical economic hubs in the region, and on the other hand, we have Cities like Shenzhen that have proliferated in only a few decades. For instance, Shenzhen has grown from a small cluster of several fishing villages that was only 1/172 of Hong Kong's GDP in 1979 to become a vibrant first-tier city in 2018 [31]. Since 2015, Shenzhen's annual population growth has exceeded 500,000, ranking first in China. The relatively orderly expansions in the PRD city cluster are somewhat expected as they were planned as part of the regional development strategies that were put in place decades ago. The opening up era and the development of China's first economic zone led to further successes that made the PRD city cluster to what it is now.

5.1.1 The Significance of the Pearl River Delta City Cluster: More Than Just a "Dual-Core" Structure

Accelerating the reform and development of the Pearl River Delta region is conducive to promoting the strategic adjustment of the economic structure of the region [24], and enhancing the impetus and vitality of economic development [4, 23]. A key aspect in the PRD region has been the development of transportation and metropolitan regions [21, 30, 62], but more importantly, the network between them. The PRD's development is conducive to improving China's comprehensive strength, international competitiveness, and ability to resist international risks [56, 59]. Its strategic location is key to its success. Therefore, the PRD development better participates in international economic cooperation and competition [5, 9, 10, 57]. It is also conducive to radiating and driving the economic development of the Pearl River Delta and Pan-Pearl River delta region [32, 37, 60] and promoting the formation of complementary advantages, positive interaction of regional economic development new pattern [33]. Politically speaking, the PRD has been developing healthily between mainland cities and the SARs. It is expected that it will continue to help implement the principle of "one country, two systems" and maintain the long-term prosperity and stability of Hong Kong and Macao [19, 22, 45]. More importantly, the PRD's development is conducive to deepening institutional innovation [7, 11–13, 18, 33], exploring new paths and providing new experiences for establishing institutional mechanisms for scientific development.

The economic and policy advantages of reform and opening up have great significance to the formation and development of the Pearl River Delta urban agglomeration [26]. In 1978, when China implemented the reform and opening up, the Pearl River Delta urban agglomeration "*ushered in a new opportunity for development*" [27], especially the reform of the forefront of Guangdong Province [28]. However, it also obtained unprecedented development. The reform of the economic system and the initial formation of the pattern of opening up greatly attracted the capital, talent, technology, and other factors of production gathered here [2, 44], paving the way for the formation of urban agglomeration in the Pearl River Delta.

The Pearl River Delta city also has some administrative regional planning advantages. The cities in the cluster belong to the same province, which is superior to the Yangtze Delta or Beijing-Tianjin-Tangshan area in the integration and coordination of resources [36]. The three provinces and cities govern the latter two, and the integration is relatively challenging. This factor can make the Pearl River Delta better in the unified planning and arrangement of the integration of urban resources, play to the advantages of each city, and mutual division of labour, which can make a virtuous circle of urban agglomeration.

Besides the administrative advantages, we note the importance of geopolitical advantages for the PRD. The advantages of the Pearl River Delta are undeniable. The PRD is adjacent to Hong Kong and Macao SARs. The initial stage of reform and opening up coincided with the upgrading of industrial structure in Hong Kong and Macau, so it needed to rely on the mainland to transfer its increasingly high cost of light products processing and manufacturing [3, 8]. Therefore, there was a large number of capital flows into the cities of the Pearl River Delta [42, 61]. Moreover, the PRD faces the South China Sea, which is very near to Southeast Asia across the sea and can be easily linked with the whole world across the ocean.

Another factor for the region's economic development success is its culture of great inclusiveness. Lingnan (Guangdong) culture does not exclude the acceptance of investors, entrepreneurs, and talents from all over the world but also fills the shortage of local resources [16]. Looking at the PRD's development process, the contribution made by the outsiders is enormous [48], and they will also play a more significant role in helping the Pearl River Delta to form a world-class city cluster. The related aspect is that it is generally known that 'enough money is flowing in' [20, 34]. Pearl River Delta is China's famous hometown of overseas Chinese, Hong Kong, and Macao compatriots, the largest number of Chinese overseas, with overseas natural convenient cultural ties [35]. Hong Kong, Macao, and overseas Chinese account for most of the foreign capital attracted by the Pearl River Delta [43], which plays a leading role in the development of an export-oriented economy.

5.2 A Micro-Historical Analysis of the Pearl River Delta City Cluster

From a historical point of view, the development of the Pearl River Delta urban agglomeration can be divided into four stages [26]. First, in the Yuan Dynasty before 1949, the founding of New China, it is evidenced that the development of the Pearl River Delta as a port region was relatively simple. Since the Portuguese occupation of Macao, Guangzhou began as a trade port and began to form a dual-center development model with Macau [55]. Consequently, the Pearl River Delta region had a gradual international trade development. After the Opium Wars, Hong Kong's economy developed rapidly, and Macao's economy declined gradually [38], which led to changes in Guangzhou's industrial structure. From a single port trade to industry and commerce, finance, transportation hub, and other comprehensive multi-functional directions.

The second stage of development was in 1949, after the founding of New China and before the reform and opening up in 1978. During this period, due to China's political repression, the mainland reduced the exchanges with Hong Kong and Macao and focused on the development of heavy industry inland. Therefore, as a coastal region, the Pearl River Delta region had an extremely slow economic development [29], basically at a standstill or even in a recession. The third stage was in 1978, when the Chinese government implemented the reform and opening up. At this time, the PRD urban agglomeration as a coastal region had a great opportunity for new economic development [27]. In particular, we can refer to the rapid development of Shenzhen's economy, Guangdong Province, Guangzhou, Shenzhen double economic center, radiation around the city, etc. The fourth and last stage (so far) of the PRD's development goes to 2015. In this year, the Guangdong Provincial Government compiled the Pearl River Delta Regional Plan (2014–2020), which established the positioning and development goals of the PRD urban agglomeration [40]. In February 2019, the "Guangdong-Hong Kong-Macao Greater Bay Area Development Planning Outline" was issued, genuinely linking the nine cities of the Pearl River Delta with Hong Kong and Macao [51]. It makes clear that the Guangdong-Hong Kong-Macau Greater Bay Area will build itself into a base for emerging industries, advanced manufacturing, and modern services worldwide. The primary aim is to build the region into a world-class city cluster and a world-class bay area.

5.2.1 An Overview of Industrial Structure in the Pearl River Delta

Apart from Shenzhen, cities in the Pearl River Delta are mainly low-end manufacturing industries. Hence, their independent innovation capacity needs to be improved, and high-quality public resources are scarce [41]. First, most cities in the Pearl River Delta are dominated by low-end manufacturing. The industrial development of the Pearl River Delta urban agglomeration benefits from the international industrial transfer in the early 1980s and early 1990s, both of which are labour-intensive and have the same industrial structure [2]. In addition to Shenzhen, the rest of the city's current pillar industries are still concentrated in machinery manufacturing, metal smelting, textile, food, chemical, and other low-end manufacturing, finance, information, new energy, new materials, and other industries slow development. Take the financial industry, for example; in 2018, the value added by the financial sector in Guangzhou and Shenzhen accounted for 12.7 and 9.1% of the total GDP, respectively [49]. This is while Beijing and Shanghai accounted for 16.8% and 17.7%, respectively. From the growth point of view, Beijing's financial sector added a value growth rate of 7.2, and 5.7%. Shanghai, Guangzhou, and Shenzhen were only 4.3 and 3.7%. Secondly, compared with the Yangtze River Delta and Beijing-Tianjin-Hebei urban agglomerations, the independent innovation ability of core cities still needs to be improved [15]. In 2018, the ratios of R&D expenditure to GDP in Guangzhou and Shenzhen were 2.8% and 4.0%, respectively, while Beijing and Shanghai reached 5.7% and 4.1%, respectively. In 2018 [54], the number of invention patents granted in Guangzhou and Shenzhen was 11,000 and 21,000, respectively, lower than that in Beijing and Shanghai, where 48,000 and 21,300 patents were granted [15].

Lastly, the shortage of quality public service resources has become a bottleneck restricting the development of urban agglomeration in the Pearl River Delta. In terms of education, there are only 4,985 and 211 universities, such as Sun Yat-Sen University and the South China University of Technology. At the same time, the Beijing-Tianjin-Hebei and Yangtze River Delta cities have more educational centers, institutes, and well-known universities [52]. In terms of medical care, the numbers of hospitals in Guangzhou and Shenzhen were 243 and 134, respectively [15], much lower than in Beijing and Shanghai.

5.3 Future Development Direction of the Pearl River Delta Urban Agglomeration

The industrial development of the Pearl River Delta has experienced the stage of accepting the transfer of Hong Kong and international processing and manufacturing industries and the growth stage of the domestic demand-oriented localisation industry [47]. At present, it is stepping into a stage of independent innovation with more complete categories and more perfect functions. The manufacturing level of the Pearl River Delta urban agglomeration is developed, and the future is committed to building science and technology, industrial innovation center and advanced manufacturing, and a modern service base [50].

Shenzhen's proximity to Hong Kong has positioned it as a hub for finance, trade and innovation in the Pearl River Delta. On the east bank of the Pearl River, the cluster includes Shenzhen, Dongguan, and Huizhou as the main body for the formation of a well-known electronic information industry corridor. On the west bank of the

Pearl River, Foshan, Zhongshan, Zhuhai, and Guangzhou are the main body for the formation of electronic and electrical machinery industry clusters [1]. Moreover, Guangzhou is not only the political, economic, and cultural center of the province but also one of the national railway and aviation hubs. Its industries are mainly automobile, electronics, and chemical industries, accounting for 56% of the total industrial output value [15].

In the new round of global scientific, technological, and industrial revolution and China's new opening up, the Greater Bay Area of Guangdong, Hong Kong, and Macao, and the Pearl River Delta are expected to build a "two main sub-three" central system. Through the "cluster of industrial clusters", the formation of regional key industries are reasonably distributed in the cluster. It is anticipated that Shenzhen and Guangzhou, the two central cities, will move from competition to cooperation and give full play to their radiation role. Shenzhen vigorously promotes technological and institutional innovation, gives full play to the leading and exemplary role of the special economic zones, and focuses on the development of high-end industries such as science and technology, finance, and foreign trade [54]. Guangzhou will comprehensively promote the new industrialisation, information technology, and internationalisation, and carefully build an economic center, a cultural city, and a landscape capital [1]. Relying on the Hong Kong-Zhuhai-Macao Bridge, the Regional Deputy Center of Zhuhai has planned to deepen the cooperation with Hong Kong and Macao, focusing on the development of trade, tourism, productive services, logistics, and other industries.

Furthermore, other cities also play a major part in this regional cooperation. Foshan emphasises developing the light industry and advanced manufacturing industry with equipment manufacturing as the core. Zhaoqing actively undertakes the Guangfo industry and capital transfer. It plans to do excellent traditional advantages industry and actively build a modern urban agricultural system and traditional advantages of industrial transformation. It plays a vital role in upgrading the cluster area and taking advantage of the regional advantages of connecting the east and the west. It also plays a hub role in transit ports, expanding the economic circle and Guangxi and Guizhou links [54]. Similarly, other smaller cities like Zhongshan and Jiangmen will strengthen comprehensive service functions, continue to strengthen cooperation with Macao, and develop international entertainment, modern business services, sightseeing, and tourism industries.

5.4 Sustainable Development of the Pearl River Delta Urban Agglomeration

The Development Goals and Tasks of Guangdong Province in the "14th Five-Year Plan" Period are to create a demonstration area for convergence of rules, a gathering place for high-end elements, a source of innovation for science and technology industries, a place for linking internal and external cycles, and a place to support security development [1]. It also plans to take the lead in exploring effective paths conducive to forming a new development pattern.

Critical tasks in 2021 include the Implementation of the national carbon peak and carbon neutral deployment requirements, promoting the peak of carbon emissions by regions and industries, and deepening the carbon trading pilot. The cluster aims to accelerate the adjustment and optimisation of the energy mix, vigorously develop clean energy such as natural gas, wind, solar and nuclear energy, and increase the proportion of natural gas in primary energy [1]. The cities of the city cluster also plan to collectively study and establish an energy budget management system and put in place strict control of new high-energy consumption projects.

5.4.1 *Challenges of Sustainable Economic Development in the Pearl River Delta Region*

There are also many problems in the process of sustainable development of the regional economy in the Pearl River Delta. Because of the unbalanced strategy of prioritising the development of coastal areas, the industry of Zhujiang Delta is growing rapidly, and the social wealth is increasing rapidly [1]. However, with the deepening of reform and opening up, especially through the implementation of the strategy of developing the western region, the balanced regional strategy has replaced the unbalanced gradient strategy in the past. Nonetheless, various contradictions and drawbacks have gradually emerged [1], mainly as follows:

(1) The shortage of technical talents leads to the low quality of economic development. Compared with other regions, the Pearl River Delta region has human resources and technology advantages. However, from the economic development perspective, there are still problems of low-quality workers, lack of professional skills, and management personnel. Hence, there is a possible chance of decline.
(2) Insufficient capital input could lead to the slow technological transformation of enterprises. Funds for technological transformation and high-tech research and development are seriously insufficient in the region. This investment situation is not beneficial to the technological transformation of enterprises but also to the innovation of core technologies, which leads to the regional economy running at the level of high input, high consumption, and low output.
(3) Talent distribution is unreasonable, which creates some contradictions in the industrial structure adjustment. The proportion of talents in the three industries is about 1:8:7, which leads to the overall imbalance of its industrial structure. On the one hand, the primary sector is relatively basic, and technology development and personnel application are insufficient. The situation affects the industrial structure's healthy development, which could weaken the overall economic development level. On the other hand, the tertiary industry is a high-quality industry. Due to the lack of technology application and talent development, the

industry content is not high and experiences a sort of slow development. Thus, there is a "double constraint" effect [1]. This is a significant obstacle to the future sustainable development of the Pearl River Delta region.

(4) The ability to coordinate economic and social development is poor, and the ecological balance is seriously damaged. Due to overemphasis on economic growth, immediate benefits, and sustainable development of the regional economy, high economic growth has become low economic growth at the expense of future generations living conditions.

(5) The level of technological development is still backward [11]. As a result, it is evidenced that the technological content of products is low, and the innovation ability of enterprises is poor. According to relevant data, only about 30% of scientific and technological achievements in the Pearl River Delta have been transformed. The contribution rate of science and technology to the economy is less than 45%, and the former is only 43 and 38% of the United States and Japan. The latter is only 56 to 75% of developed countries.

5.4.2 Summary: Sustainable Development Responses to Carbon Neutrality Plan

There are four main sustainable development responses to the carbon neutrality plan. The first is to conclude government cooperation agreements between members of economic regions to achieve win–win cooperation. Under the condition that the market economy has not developed to a high degree, it is not enough to break through the barriers of the administrative region economy by relying solely on the market's own integration mechanism. The regional coordination mechanism must be established through horizontal administrative and institutional cooperation among local governments to balance the tension between economic integration and local interest separation. The second point is to increase personnel training and introduction. The upgrading and integration of regional education resources, especially the promotion of higher education, plays a decisive role in the sustainable development of the whole Pearl River Delta regional economy. The introduction of external talents is a critical way to solve the shortage of talents.

The third response is to majorly transform the regional economic development model. The mode of regional economic development should be changed from labour-intensive to knowledge-intensive to improve regional products' competitiveness and promote economic development. Lastly, a major response is to have consolidated capital at the regional level. It is vital that the response focus on integrating the capital in this region, giving space to the role of capital in industrial upgrading, and strengthening cooperation in the capital market.

5.4.3 Case Study Example: The Greater Bay Area

An ongoing case study example is the Greater Bay Area (Guangdong-Hong Kong-Macao Greater Bay Area or GBA). The area includes Hong Kong SAR, Macao SAR, and nine cities of the Pearl River Delta. Building the Guangdong-Hong Kong-Macao Greater Bay Area is a new attempt in the new era to promote a new pattern of all-around opening up and new practice in advancing the cause of "one country, two systems". On February 18, 2019, the CPC Central Committee and the State Council issued the Outline of the Development Plan for the Guangdong-Hong Kong-Macau Greater Bay Area [1]. According to the plan, the Guangdong-Hong Kong-Macau Greater Bay Area will not only be built into a vibrant world-class city cluster. Still, it will offer more sustainable development strategies, carbon neutrality initiatives, sustainable economic development, etc. It is planned to include an international science and technology innovation center, important support for the Belt and Road Initiative, and a demonstration zone for in-depth cooperation between the mainland, Hong Kong, and Macao. But it is also vital to create a livable industry that could be suitable for tourism quality life circle. This regional development could become a model of high-quality development. Hong Kong, Macao, Guangzhou, and Shenzhen, the four major cities as the core engine of regional development, play a major part in enhancing cooperation between cities and in the region.

It is of great significance for China to better implement the strategic policy of "one country, two systems", which could benefit the region even beyond China's borders. The construction of the Guangdong-Hong Kong-Macao, Greater Bay Area, is conducive to China's better implementation of "one country, two systems." The greater bay area of Guangdong, Hong Kong, and Macao will provide greater organisational cooperation for implementing the Closer Economic Partnership Arrangement (CEPA), which was initially proposed at the beginning of 2004. The advantage of geographical proximity is conducive to enhancing the depth and breadth of cooperation between Hong Kong, Macao, and the Mainland cities. Consequently, it will support the sustained economic prosperity of Hong Kong-Macao.

The construction of the Guangdong-Hong Kong-Macao Greater Bay Area, is also conducive to developing China's central and western regions and implementing the west development strategy. This strategy has not succeeded previously, and the new plans could help gradual sustainable development of those regions. A major shift is expected to strengthen the economic cooperation between the great Pearl River Delta and the surrounding backward areas. It is anticipated that the pattern of complementing each other's advantages and developing together could emerge into a model of sustainable regional development. Therefore, it will be conducive to the implementation of the strategy for the great development of the western region.

5.5 Pathways to Implementing and Achieving the National-Level Carbon Neutrality Plan

The PRD's plan to achieve carbon neutrality depends on major industrial restructuring at the cluster level. Therefore, it is essential to consider the low carbonisation of the energy system [14] to improve the operation flexibility and promote the energy mix plan through better integration opportunities [46]. Also, findings from the study by [53] suggest the role of "*emission reduction policies to achieve carbon neutrality*…[which]…*could curb air pollution but would contribute to higher air temperature*". Thus, in a sub-tropical warm region like the PRD, this could lead to other challenges like climate resilience, cooling, and climate change mitigation. [53] also argue that "*the Ambitious-Pollution-Neutral-Goals scenario from the Dynamic Projection model for Emissions in China (DPEC) database*" could be an example of china's future emission reduction pathway, particularly in the regions like the PRD.

In addition, the Pearl River Delta city cluster needs to focus on high-quality growth, comprehensive enough to include better economic growth, better integration of technologies, reduced energy consumption and low added-value development, and focus on industrial transformation. The region's manufacturing hubs require to go through transitional phases of decarbonisation to ensure they follow the process of industrial restructuring that is already in process. As mentioned earlier, the city cluster should also help achieve a higher level of diversity in production and economic development, which could then allow for implementing a better energy mix for the regional industries. The PRD has already gone beyond the dual-core structure system to a city cluster economic development approach. In fact, this city cluster is one of the first city clusters that boosted regional development in the face of regional competitiveness and dual structure systems. Therefore, we anticipate a better investment in carbon storage strategies, biomass carbon stock, and rural–urban cooperation for integrating renewable energies, better waste management systems, sustainable transportation, etc. The connectivity in this city cluster is the backbone to achieving these targets in a holistic pathway to carbon neutrality.

References

1. Baike (2020). Outline of the development of the Pearl River Delta <珠三角发展纲要> Available from: https://baike.so.com/doc/6947054-7169420.html
2. Bie J, Jong MD, Derudder B (2015) Greater Pearl River Delta: historical evolution towards a global city-region. J Urban Technol 22(2):103–123
3. Chan RC (2011) Hong Kong's position in Guangdong under' one country, two systems'. Seoul J Econ 24(2):125–150
4. Chan RC, Yao S (2010) New trends in urban and regional governance in the Pearl River Delta region. Asian Geogr 27(1–2):127–143
5. Chen B, Ma J, Feiock R, Suo L (2019) Factors influencing participation in bilateral interprovincial agreements: evidence from China's pan pearl river delta. Urban Aff Rev 55(3):923–949

6. Chen L, Mu T, Li X, Dong J (2022) Population prediction of Chinese prefecture-level cities based on multiple models. Sustainability 14(8):4844.z
7. Chen Q, Wang CH, Huang SZ (2020) Effects of organizational innovation and technological innovation capabilities on firm performance: evidence from firms in China's Pearl River Delta. Asia Pac Bus Rev 26(1):72–96
8. Cheng JY, Yeung YM, Chu DK (eds) (1998) Guangdong: survey of a province undergoing rapid change. Chinese University Press, Beijing
9. Cheshmehzangi A (2016) China's New-Type Urbanisation Plan (NUP) and the foreseeing challenges for decarbonization of cities: a review. Energy Procedia 104:146–152
10. Cheshmehzangi A (2020) Towards urban energy management: a brief review of China's recent transition in urban energy demand. Low Carbon Econ 11(01):1
11. Cheshmehzangi A, Chen H (2021a) China's sustainability transitions: low carbon and climate-resilient plan for carbon neutral 2060. Springer Nature, Singapore
12. Cheshmehzangi A, Chen H (2021b) Overview of China's low carbon progress: policies and initiatives. In: China's sustainability transitions. Springer, Singapore, pp 103–117
13. Cheshmehzangi A, Dawodu A, Sharifi A (2021) Sustainable urbanism in China. Routledge, New York
14. Cheshmehzangi A, Li H, Yang X (2018) Low carbon and climate resilient urban development in China. ADB Report, developed for the National Development Reform Committee (NDRC), Beijing, China
15. Chi-tech Industry Research Institute (2020) Industrial Map of China's five urban agglomeration (2) Pearl River Delta city agglomeration. <中国五大城市群产业地图之(二)珠三角城市群>. Available from: https://www.sohu.com/a/387632886_423490
16. Chongyi F (1999) Seeking lost codes in the wilderness: The search for a Hainanese culture. China Q 160:1036–1056
17. Fei W, Zhao S (2019) Urban land expansion in China's six megacities from 1978 to 2015. Sci Total Environ 664:60–71
18. Fu W (2015) Towards a dynamic regional innovation system: Investigation into the electronics industry in the Pearl River delta, China. Springer, Cham
19. Geping R, Zhenmin W (2007) Hong Kong's 'One Country, Two Systems' experience under the basic law: two perspectives from Chinese legal scholars. J Contemp China 16(52):341–358
20. He X (2009) Enforcing commercial judgments in the Pearl River Delta of China. Am J Comp Law 57(2):419–456
21. Hou Q, Li SM (2011) Transport infrastructure development and changing spatial accessibility in the Greater Pearl River Delta, China, 1990–2020. J Transp Geogr 19(6):1350–1360
22. Leng T (2011) On the fundamental characteristics of the "one country, two systems" policy. Acad J One Ctry, Two Syst 1:49–59
23. Li F, Xing W, Su M, Xu J (2021) The evolution of China's marine economic policy and the labor productivity growth momentum of marine economy and its three economic industries. Mar Policy 134:104777
24. Li L, Li Y, Ye F, Zhang L (2018) Carbon dioxide emissions quotas allocation in the Pearl River Delta region: evidence from the maximum deviation method. J Clean Prod 177:207–217
25. Li X, Tan Y, Xue D (2022) From world factory to global city-region: the dynamics of manufacturing in the Pearl River Delta and its spatial pattern in the 21st century. Land 11(5):625
26. Li Y, Chen Y (2021) Development of an SBM-ML model for the measurement of green total factor productivity: the case of pearl river delta urban agglomeration. Renew Sustain Energy Rev 145:111131
27. Liang L, Chen M, Luo X, Xian Y (2021) Changes pattern in the population and economic gravity centers since the Reform and Opening up in China: the widening gaps between the South and North. J Clean Prod 310:127379
28. Liang Y, Zhou Z, Li X (2019) Dynamic of regional planning and sustainable development in the Pearl River Delta, China. Sustainability 11(21):6074
29. Lin GC (1997) Red capitalism in South China: growth and development of the Pearl River Delta, vol 2. UBC Press

30. Lin GC (1999) Transportation and metropolitan development in China's Pearl River Delta: the experience of Panyu. Habitat Int 23(2):249–270
31. Liu AY (2018) Building markets within authoritarian institutions: the political economy of banking development in China. Stanford University
32. Liu T, Li Y (2021) Green development of China's Pan-Pearl River Delta mega-urban agglomeration. Sci Rep 11(1):1–11
33. Liu X, Wang Z, Cui X (2021) Scenario simulation of the impact of China's free-trade zone construction on regional sustainable development: a case study of the Pearl River Delta urban agglomeration. Sustainability 13(14):8083
34. Lo VH, Yeung AH (2004) Practical framework for strategic alliance in Pearl River Delta manufacturing supply chain: a total quality approach. Int J Prod Econ 87(3):231–240
35. Lu J (2012) A Social-network behavioural approach to overseas Chinese and overseas non-Chinese investments in China. Tijdschr Econ Soc Geogr 103(4):426–442
36. Ma Y (2008) Comparison of advantages and weaknesses among three major urban agglomerations in China. Asian Soc Sci 4(10):132–134
37. Mei N (2019) Comparison of population and industrial structure between urban agglomerations of the Pearl River Delta and world-class urban agglomerations. Mod Econ 10(03):886
38. Meyer DR (2000) Hong Kong as a global metropolis, vol 30. Cambridge University Press, Cambridge
39. Ouyang X, Mao X, Sun C, Du K (2019) Industrial energy efficiency and driving forces behind efficiency improvement: evidence from the Pearl River Delta urban agglomeration in China. J Clean Prod 220:899–909
40. Shen L, Cheng G, Du X, Meng C, Ren Y, Wang J (2022) Can urban agglomeration bring "1+ 1> 2Effect"? a perspective of land resource carrying capacity. Land Use Policy 117:106094
41. Sheng Y, Zhao J, Zhang X, Song J, Miao Y (2019) Innovation efficiency and spatial spillover in urban agglomerations: a case of the Beijing-Tianjin-Hebei, the Yangtze River Delta, and the Pearl River Delta. Growth Chang 50(4):1280–1310
42. Sit VFS (1991) Transnational capital flows and urbanization in Pearl River Delta, China. Asian J Soc Sci 19(1):154–179
43. Sit VF, Yang C (1997).Foreign-investment-induced exo-urbanisation in the Pearl River Delta, China. Urban Stud 34(4):647–677
44. Smart A, Hsu JY (2004) The Chinese diaspora, foreign investment and economic development in China. Rev Int Aff 3(4):544–566
45. So AY (2011) "One country, two systems" and Hong Kong-China national integration: a crisis-transformation perspective. J Contemp Asia 41(1):99–116
46. Southeast University (2020) SEU-Yangtze River delta carbon neutrality strategy development institute launched, a report summarised by Southeast University School of Energy and Environment. Available from: https://power.seu.edu.cn/ny_en/2020/1228/c24915a357896/pagem.htm
47. Sun L, Qin L, Taghizadeh-Hesary F, Zhang J, Mohsin M, Chaudhry IS (2020) Analyzing carbon emission transfer network structure among provinces in China: new evidence from social network analysis. Environ Sci Pollut Res 27(18):23281–23300
48. Tang Y (2012) World Factory in crisis: global economy downturn as a speedbump on the path of development in China's Pearl River Delta, and the twisty road for recovery. Available from: https://www.theseus.fi/handle/10024/53241
49. Di Tommaso MR, Rubini L (2013) Southern China: industry, development and industrial policy, vol. 87. Routledge
50. Wang MX, Zhao HH, Cui JX, Fan D, Lv B, Wang G, Li ZH, Zhou GJ (2018) Evaluating green development level of nine cities within the Pearl River Delta, China. J Clean Prod 174:315–323
51. Wu J (2022) A study on the adaptability of art majors in guangdong undergraduate colleges under the background of the planning and development of the Guangdong-Hong Kong-Macao Greater Bay Area. In: Knowledge innovation on design and culture: proceedings of the 3rd IEEE international conference on knowledge innovation and invention 2020 (IEEE ICKII 2020), pp 41–43

52. Sha X, Power Grid Headlines (2021) How to peak and carbon neutral? inventory 23 provincial administrative regions to develop programs. 碳达峰、碳中和怎么干?盘点23个省级行政区域制定方案. Available from: http://www.tanpaifang.com/tanzhonghe/2021/0521/77905_20.html
53. Xu B, Wang T, Ma D, Song R, Zhang M, Gao L, Li S, Zhuang B, Li M, Xie M (2022) Impacts of regional emission reduction and global climate change on air quality and temperature to attain carbon neutrality in China. Atmos Res 106384 (in press)
54. XZBU (n.d.) A brief analysis of the sustainable development of the pearl river delta regional economy 浅析珠三角区域经济的可持续发展. Available from: https://www.xzbu.com/3/view-11536417.htm
55. Yan Y, Ju H, Zhang S, Jiang W (2019) Spatiotemporal patterns and driving forces of urban expansion in coastal areas: a study on urban agglomeration in the Pearl River Delta, China. Sustainability 12(1):191
56. Yang C (2012) Restructuring the export-oriented industrialization in the Pearl River Delta, China: institutional evolution and emerging tension. Appl Geogr 32(1):143–157
57. Yang C, Li SM (2013) Transformation of cross-boundary governance in the Greater Pearl River Delta, China: contested geopolitics and emerging conflicts. Habitat Int 40:25–34
58. Yang Y, Qu S, Cai B, Liang S, Wang Z, Wang J, Xu M (2020) Mapping global carbon footprint in China. Nat Commun 11(1):1–8
59. Zeng DZ (2015) Global experiences with special economic zones: focus on China and Africa. World Bank Policy Res Work Pap (7240).
60. Zhang W (2022) The influence of Hong Kong-Zhuhai-Macao Bridge on the development of urban agglomeration in the Pearl River Delta. Secur Commun Netw
61. Zhang X, Chen S, Luan X, Yuan M (2021) Understanding China's city-regionalization: spatial structure and relationships between functional and institutional spaces in the Pearl River Delta. Urban Geogr 42(3):312–339
62. Zheng S (2019) Has "The outline of the plan for the reform and development of the Pearl River Delta" promoted to the coordination of Guangdong-Hong Kong-Macao Greater Bay Area. Mod Econ 10(4):1348–1367

Chapter 6
Chengdu-Chongqing City Cluster: From an Economic Zone to a Major Economic Circle in the Western China

6.1 A General Introduction to Chengdu-Chongqing City Cluster

The Chengdu-Chongqing city cluster is recognised as a major economic hub, mostly known to be a Twin City Region in the West of China. This is the fourth main economic cluster in the country, putting together a significant economic hub in the region. It is located in the southwest of China, the combination of east and west, the intersection of north and south, east of Hunan and Hubei, west of Qinghai-Tibet, south of Lianyungui, and North of Shaanxi and Gansu. This main city cluster is to undertake South China and Central China, connecting Southwest and Northwest, enhancing the central and western communication, and linking to the south of the country and neighbouring countries/regions in Southeast Asia[37]. Thus, it is identified as an important intersection and transportation corridor.

Historically, the region has been an important node in the western part of China, one of the very few western regions with major economic hubs. More specifically, the newly proposed Chengdu-Chongqing city cluster is centered on the two core cities of Chengdu and Chongqing, along the upper regions of the Yangtze River, the Baozhou-Chengdu-Kunming Railway, and the regional expressway that connects the region more extensively[37]. Based on the earlier reports, Chengdu-Chongqing Railway and Expressway, Chengsui Chongqing Railway, Expressway, and Chengnan Expressway are the main transportation axes that have already helped with enhanced enhanced enhanced transportation connectivity, rapid trade, and business mobility, etc. The connections go out of the Three Gorges in the east, into Hunan and Hubei province (also in the east), and down the Yangtze River, directly connecting with the developed Changjiang River Delta. Geographically speaking, the west of the Qinghai-Tibet Plateau becomes the inland gateway into Tibet and the rear base of Tibet autonomous region. The other prominent geographical advantages are connections to the Yunnan-Guizhou Plateau, the Southeast to the Pearl River Delta, the South China Sea, etc. Also, towards the north, the region benefits from the rich

A. Cheshmehzangi and T. Tang, *China's City Cluster Development in the Race to Carbon Neutrality*, Urban Sustainability, https://doi.org/10.1007/978-981-19-7673-5_6

minerals and natural environments of Qinba mountain, then Shaanxi, Gansu, and Ningxia Qingxin gas hubs.

Although smaller than some other city clusters, the Chengdu-Chongqing cluster covers a major area of cities, towns, and villages. The city cluster includes the areas in Chongqing, including all its urban districts and countries (i.e., Fengdu County, Dianjiang County, and Zhong County). The cluster also covers a larger area of 15 cities in Sichuan Province, including Chengdu, Zigong, Luzhou, Deyang, Mianyang (except Beichuan County and Pingwu County), Suining, Neijiang, Leshan, Nanchong, Meishan, Yibin, Guang'an, Dazhou (except for Wanyuan City), and Ya'an (excluding Tianquan and Baoxing). In a vast area of 185,000 square kilometers, the Chengdu-Chongqing city cluster is a populated agglomeration with an estimated resident population of 100 million in 2020. As per recent records, the city cluster has a total GDP of 6.8 trillion yuan (approximately 1.07 trillion USD—as of December 2021), accounting for 6.7% of China's GDP.

6.1.1 The Cluster's Significance: A Twin City Region and Economic Circle

Due to its economic importance, the Chengdu-Chongqing city cluster is identified as a world-class city cluster [17]. The iChongqing news coverage of the cluster (ibid) refers to the poem of Li Bai from the Tang Dynasty (618–907 A.D.) that says "*the hard roads all toward Shu*"—which refers to the ancient name of Chongqing and Sichuan areas (i.e., Shu). Hence, the region's economic importance goes way back to much earlier years in China's history. The reports of the National Development and Reform Commission (NDRC) suggest the cluster as the "New International Land-Sea Trade Corridor", which was released in August 2020 (ibid). The larger masterplan is aimed to develop a major hub for logistics and operation organisation center, covering not only the region but also the corridors and nearby hubs connecting to it. The reliance of the economic circle on the main corridor is suggested to be developing the region into a significant passage economy.

The superior inland location makes it convenient for the Chengdu-Chongqing city cluster to allocate resources and develop markets in a larger space. As a major advantage of connecting multiple regions and the only major economic circle of Western China, the Chengdu-Chongqing city cluster significantly enhances regional economic prosperity. Much of this is owed to its logistics and trade connections, as well as those proposed to link the region to neighbouring sub-regions and countries in the South-East Asian region.

Chengdu-Chongqing city cluster as the east–west, north–south intersection node, and corridor is conducive to the regional planning of the resources of South China, Central China, and Southwest Northwest. Moreover, it is believed to be conducive to enhancing the city cluster elements' influence capacity to a larger extent of the country. It also has a significant influence on a vast inland market.

As the upper reaches of the Yangtze River, the Chengdu-Chongqing city cluster is already promoting the construction of China's Yangtze economic belt, which will be conducive to actively undertaking the eastern coastal element transfer along the Changjiang River basin. This economic circle promotes Yangtze's upper and middle reaches and the downstream areas of the resources element orderly flow and industrial gradient coordinated development. In addition, Chengdu-Chongqing city cluster plays a vital role in the communication between Central Asia, South Asia, and Southeast Asia in China, which is conducive to the regional radiation and influence the factor market for Central, Southern, and Southeastern Asia [37].

Some of the shift in this city cluster is also focused on integrating science and technology transfer services, deepening the construction of technology transfer platforms, and accelerating the region's economic prosperity. A large concentration of new institutional development is on the development of new technology transfer institutions, highlighting the provisions of "science and technology policies, physical carriers, and services in support" [17]. As per CGCN (2020) reports, the leadership in the Communist Party of China (CPC) aims to bring the region closer together to bring growth to this less-developed western region of the country. The proposed dual circulation strategy seeks to be a longer-term regional development to promote technology or technological innovation and high-quality growth beyond just the city cluster's boundaries. Current economic reports indicate a faster acceleration of new investment and development in this city cluster, particularly from the mid-2020s. In May 2020, President Xi of China announced the idea of "dual circulation", which was later linked to the national plan to promote internal circulation. This approach is also related to the dynamics of domestic production, consumption, and distribution cycles, helping to upgrade the economy and innovation for future sustainable development plans and economic growth. The proposed internal circulation plan is primarily supported by external circulation, mainly linking with China's Belt and Road Initiative (BRI)[4, 5, 7, 8]. While the struggles remain for opening-up national policy, it is evident that reforms and competitiveness will remain as critical strategies for regional development, particularly for promoting innovation and sustainability plans in regions like the Chengdu-Chongqing city cluster.

The new city cluster may be new only in name, but in fact has been developing much earlier as an economic zone, and specifically for industrial collaboration [48], urban–rural development [32], transportation and accessibility, and socio-economic indicators [70]. In more recent years, we see more shifts towards the development of sustainability plans, socio-spatial considerations, and social networks [42], leading to land-use transitions [38] and towards enhanced regional coordination and cooperation. In this regard, we see the progress to be quite tangible in the economic sustainability and economic growth directions, which may require further attention when linking to achieving carbon neutrality plan and the subsequent sustainable development agenda of the region.

6.2 A Micro-Historical Analysis of the Chengdu-Chongqing City Cluster

Historically, since the founding of the People's Republic of China in 1949, Chengdu and Chongqing have been two major cities in Sichuan and the western regions of China. Since 1997 and the establishment of Chongqing municipality, Chongqing and Chengdu administrative divisions have separated. However, the two places have promoted their cooperation which has been ongoing since then. Subsequently, in 2001, the two sides signed the "Chongqing-Chengdu economic cooperation talks summary", proposed to work together to build "Chengdu-Chongqing economic corridor" [13, 35, 56, 63]. In 2003, the Chinese Academy of Sciences research report "*Preliminary Study on Key Regional Planning for the Development of China's Western Region*" proposed that in a 10-years plan. In their plan, they urged for the need to actively build the largest dual-core urban cluster in western China, with Chengdu and Chongqing as the center and central cities at all levels connecting and cooperating with each other. An earlier paper also highlighted the rationale for such development and discussed the implementation and prospects of this forthcoming plan [25]. Later studies focus on the importance of this twin city economic core strategy on carbon emission reductions [50] and towards achieving carbon neutrality through industrial restricting, logistics/transportation, and regional cooperation [4, 5]. Forming as the largest strategic support for China's western development, this city cluster is now aimed to promote the western region population, industry, information, science and technology, and culture gathering center [53]. The core of the upper reaches of the Yangtze River economic belt, making it more significant at a much larger scale. In 2003, it was the first time the Chengdu-Chongqing economic zone concept appeared in the national level report [20, 31]. Soon after, the idea merged into developing a major economic circle and towards regional development.

In April 2016, the National Development and Reform Commission (NDRC) and the Ministry of Housing and Urban–Rural Development (MoHURD) jointly issued the Chengdu and Chongqing Urban Agglomeration Development Plan [4, 26, 68], defining the construction of a national city cluster plan, leading the development and opening-up of the western region [72]. According to [30], the development orientation of Chengdu and Chongqing is as follows: (1) the critical modern industrial base in China, (2) the innovation-driven pilot area in the west, (3) the strategic highland of the inland open economy, (4) the demonstration area of urban and rural development, and (5) the leading area of beautiful China.

6.2.1 An Overview of Demographic Patterns

According to the data [1], Chengdu-Chongqing urban agglomeration has become the new population flow area in China's four major city clusters. The population of Chongqing in 2020 has increased significantly compared to 2010, and its share in the country has risen by 0.12% points. Although the proportion of the national population in Sichuan declined slightly from the population trend of Sichuan itself, the recent 20 years of population trends showed an apparent V-shaped growth. The permanent population of Sichuan was 82.35 million in 2000, 80.41 million in 2010, and 83.67 million in 2020. The steady growth is relatively smaller than the other high-level city clusters in China but still indicates the region's substantial population growth. In the first ten years, the population of Sichuan lost 2 million people, and in the second ten years, the population increased by 3 million people. The change has mainly been based on recent investment in economic development and the development of the economic circle in between two major cities. The Chongqing data also shows a V-shaped growth trend. In 2000, the resident population of Chongqing was 30.51 million, dropped to 28.85 million in 2010, and rose again to 32.05 million in 2020. Chongqing's population has increased both in size and in the country. Thus, we see the trend to be very similar in the region.

From the historical population data, 2011 is the turning point of the population trend in the Chengdu-Chongqing region [17]. In 2011, Sichuan's resident population increased by 80,000 from the previous year. The resident population changed from a decrease to an increase. Since then, it has been maintained, and the increase is gradually increasing [1]. The earlier drip suggests a specific node regarding the decline in population growth and economic development. This node and data reflect the international financial crisis. Thus, we see that the Chengdu-Chongqing region undertook a large-scale industrial transfer brought about by the population flow changes of recent development. In the last 10 years, it is also identified that the Chengdu-Chongqing region is one of the areas with the fastest economic growth in China. Moreover, under the new regional development strategy of the country, the trend of large numbers of population flowing to the Chengdu-Chongqing area in the future may be more obvious. Currently, the data suggests a steady growth and more regional development plans in the future direction. Therefore, the proposed coordination and cooperation suggest an important direction towards economic development and growth in the region. According to [59], there are tangible correlations between transportation, regional economy, and demographic space of the region. Thus, the current decade of the 2020s play a significant part in developing the region into a major economic hub in the west of China.

6.2.2 *An Overview of Industrial Structure in the Chengdu-Chongqing City Cluster*

Chengdu-Chongqing Economic Zone has formed a relatively complete industrial system, which has a more significant influence in the country's enterprises and industries on machinery, electronics, chemicals, metallurgy, military, energy, medicine, food, and beverage. According to the sales revenue ranking of 39 industries, 10 industries in Sichuan enter the top 10 in China. Chongqing has two of these top industries, and from the industry breakdown, Sichuan sales revenue into the top 10 (including the 10th) industry. Overall, Sichuan has 47 major companies and enterprises associated with these top-ranking industries, and Chongqing has 11. The industry structure in this economic circle has formed a distribution of four areas. "Four" that is to (1) Chongqing as the center of the automobile, motorcycle, electronic information, electrical equipment, engineering machinery, instrumentation, medicine, trade, logistics, and other industrial distribution; (2) Chengdu as the center of major equipment, engineering machinery, electronic information, bio-pharmaceutical, food, textile, new materials, agriculture, finance, trade, logistics, and other industrial distribution; (3) South Sichuan city group as the main body of salt chemical, natural gas chemical, phosphorus and phosphorus chemical, energy, building materials industry distribution; and (4) The distribution of natural gas, light textile, building materials, agriculture and animal husbandry/agriculture, as well as the processing of agricultural and sideline products, is mainly in central, northern and eastern Sichuan. Earlier studies suggest the importance of regional industrial structuring in the Chengdu-Chongqing Economic Zone before it was developed into a city cluster plan [21, 73].

A primary belt to ensure industrial cooperation supports the city cluster and coordination is maintained in the region [34]. "The Belt" is a high-tech industrial belt with major technical equipment, electronic information, biotechnology, medicine, new materials, finance, and logistics, formed along the Chengdu-Mianyang railway and expressway [51]. The existing industrial distribution pattern of "Four Areas and One Belt" has five characteristics, namely (1) industry highly concentrated in the central city, (2) relying on the distribution of resources, (3) industrial parks become high-tech enterprises gathering area, (4) specialised cooperation is not strong, and (5) system division. Thus, we also provide an analysis on the problems of industrial structure in the Chengdu-Chongqing city clusters, or economic zone, which are identified to be threefold.

(1) ***Large regional differences in economic levels***
Based on the current situation of the city cluster, it is evident that regional economic and spatial distributions are uneven. Thus, we see that the regional economic level gap is more considerable. Overall, the city cluster represents the main urban areas of two major cities in the region, namely Chongqing and Chengdu, as the centers to the surrounding ink stains gradient expansion pattern. The Chengdu plain and southern Sichuan are intensive economic zones in Sichuan province, with a high degree of economic concentration [54, 61, 65]. In particular, with an area of about one-tenth of Sichuan's area under its jurisdiction, Chengdu plain gathers nearly half of the province's total economic output [47]. The northeast region of Sichuan, located on the eastern edge of the Sichuan basin, is the most backward region in Sichuan's economic development with low potential energy and low economic density. This area accounts for less than one-fifth of the province's total economic output in more than half of its area of jurisdiction. Also, in this city cluster, the main urban area of Chongqing is the area with the highest level of economic development and the strongest gathering ability in Chongqing. However, the two wings of Chongqing's "One Circle

and Two Wings" [69]—Chongqing Northeast Economic Zone (Three Gorges Reservoir Area) [3] and Southeast Chongqing (Guizhou and Yunnan Provinces) [22]—have low economic density and fragile ecology [28, 66], which belong to the low potential energy areas of Chengdu-Chongqing Economy Zone.

(2) ***Serious tendency of industrial structure***
The function system of cities and towns in the Chengdu-Chongqing Economic Zone is unclear [55]. Hence, the mechanism of coordinating regional development is not perfect, and there are problems of industry convergence [13, 60] and redundant construction [42, 52]. The industrial structure in the region is somewhat similar. Thus, the industrial level is low, enterprises are mainly engaged in rough resource processing, and there are fewer enterprises for deep processing and finishing. Consequently, there are more labor-intensive enterprises, their products are of low grade, and their structuring is slowly upgraded. At the same time, the production and processing equipment is relatively backward, with low technical level, high raw materials, and energy consumption [57]. In addition, other factors have hindered economic cooperation and industrial contacts between Sichuan and Chongqing. These factors include backward enterprise operation and management [12], lack of a modern enterprise management system [41], the slow turnover of funds, high management costs [2], and so on. In the final analysis, the reason lies in the extensive mode of production. The Central Economic Work Conference at the end of 2009 profoundly expounded the importance and urgency of changing the pattern of economic development [9], emphasising that in the current economic situation, the transformation of the pattern has become urgent [57]. For Chengdu-Chongqing Economic Zone, a profound understanding of the current economic zones, domestic and international economic development situation could help accelerate the transformation of economic development mode as an urgent matter.

(3) ***Inter-industry linkages are relatively loose***
Chengdu-Chongqing Economic Zone is the region with the best economic foundation and the greatest development potential in the western parts of China. It is also a key development zone in the state's leading functional area planning and western development planning [24]. Sichuan and Chongqing are similar in natural geography, humanity, and customs and have frequent social and economic exchanges [39]. The market relies on each other, the industrial correlation and complementarity are strong and have a good cooperation foundation [62]. However, the degree of specialisation and cooperation in the region is low, and the industrial chain has not yet formed [49]. In recent years, through large-scale readjustment of the industrial layout, the concentration of industrial enterprises has greatly improved [40]. Because there are many large central enterprises in the region, there are also many state-owned enterprises and national defense science and technology enterprises. Thus, under the division of administrative subordination, the central industrial enterprises, defense science, technology, industrial enterprises, and central and local enterprises and defense and civilian enterprises have their own restrictions. As a result, there are few horizontal links between enterprises. This hugely affects the industrial

change, which is short and weak. To summarise, it is evident that it is difficult to embrace the benefits of aggregation fully. Thus, further adjustments are needed in the future steps of industrial restructuring processes, coordination, and management.

6.3 Sustainable Development Plan: An Economic Zone with the Ecological Civilization Agenda

The existing reports indicate countermeasures for optimising industrial structure in the Chengdu-Chongqing city cluster [29, 55, 58]. Therese countermeasures are twofold, as summarised below:

(1) **Promote the division and cooperation of industries in Chengdu-Chongqing Economic Zone**
According to the principle of complementary advantages and optimal allocation, establishing a closer industrial division of labour and cooperation is the primary task of the future industrial development of the Chengdu-Chongqing Economic Zone [19, 74]. In the industrial division of labour and cooperation in the Chengdu-Chongqing economic zone, attention should be paid to giving play to their respective comparative advantages and forming a complementary industrial relationship through the flow and reorganisation of production factors [10, 36, 38, 43]. From the existing division of labor structure, the emphasis should be placed on the industrial cooperation between the cities in Sichuan Administrative Region and the districts and counties in Chongqing Administration Region [51]. It is necessary to accelerate the readjustment of the industrial structure of the two places and strengthen bilateral cooperation in such areas as electronic information, machinery, high and new technology, biomedicine, tourism, logistics, and finance [11]. In terms of industrial development, the cities, districts, and counties in this city cluster should establish a relationship of the division of labour and cooperation between the upstream and downstream industries [51] and between horizontal industries, according to the local development orientation.

(2) **Vigorously develop industries in line with the concept of ecological civilization**
Compared with other economic zones in the country, Chengdu-Chongqing Economic Zone accounts for a large proportion of the primary industry. Future development faces the enormous pressure of upgrading and optimising the industrial structure [27, 67]. In the following steps, the Chengdu-Chongqing Economic Zone should focus on the goal of forming and developing the industry of "*saving energy resources and protecting ecological environment*," and give priority to developing the common superior industries in Sichuan and Chongqing, and realise the clustering of the common advantageous industries [51]. The approach should accelerate the optimisation and upgrading of existing industrial clusters. The results from such a sustainability transition should then

cultivate high-end clusters and play a leading role in large enterprises and well-known brands. This direction, as highlighted by [51], should accelerate the readjustment of the organisational structure of large state-owned enterprises and gradually form a new pattern of joint development of large, medium, and small enterprises, with large-scale enterprises with advantages as the core [14, 51, 57, 68]. At the same time, scientific research will jointly speed up the development of high-tech industries [18, 51, 64] and implement a series of supporting measures.

The establishment of "saving energy resources and protecting the ecological environment" industrial structure helps the formation of ecological civilization construction of industrial structure system [5, 8, 15]. How to understand and construct this system can be understood and processed from three aspects: first, considering from the system function, under the background of industrial civilization, we separate the product production and waste disposal, and the product producer and the subject of environmental protection have not been unified. Therefore, establishing "saving energy resources and protecting the ecological environment" of the industrial structure requires these factors to be unified into the industrial system [51]. Secondly, considering the relationship between the three industries, the industrial structure of "saving energy resources and protecting the ecological environment" is an industrial system composed of ecological agriculture, ecological industry, and ecological service industry and their interaction and coupling (ibid). Third, from the point of view of the production system, the industrial structure of "saving energy resources and protecting the ecological environment" is an industrial system of the ecological production process and ecological products (ibid). Thus, the transitions are likely to strengthen the ecological civilization plan, if cooperation and coordination are strengthened beyond just the twin-city arrangement and more of the regional-scale development.

6.4 Summary of the Pathways Towards Achieving China's Carbon Neutrality Plan (CNP)

In the Chengdu-Chongqing city cluster, cities play a significant part in carbon peaking and achieving carbon neutrality targets [71]. There are two groups of cities, including the core cities of Chengdu and Chongqing, and other cities of various sizes, such as medium and smaller cities of the region. Based on the current situation of this city cluster, a much stronger network is needed to strengthen the spatial association and CO_2 emission monitory of the region [42]. The larger-scale emphasis on transportation and logistics development suggests a vital driver of transit-oriented development and networked cities and towns in the region. Therefore, the current situation should help strengthen rural–urban cooperation and better coordination between multiple industries of the region. As a significant economic zone of Western China, the focus can rely on developing a robust economic circle and emphasise or circular economy and regional coordination for carbon emission reductions, production, distribution, and consumption patterns. A good example is a recent direction of implementing air pollution policies [13], using incentives of low-level consumption to promote CO_2 emission reductions, and supporting industries to restructure through a transitional phase.

In this cluster, the foremost opportunity to meet the CNP target is the regional coordination of energy-incentive industries and logistics planning policies. Suppose the industrial chain management succeeds by the end of 2030, completing the carbon peak time. In that case, the earlier ideas on integration and sustainable development of the cluster could become a reality. Thus, the earlier suggestions could be valid to help sustainability transitions of "industries, population and spatial integrations, and the function" (ibid) and towards growing a mature industrial structure at the regional level. Yet, the earlier countermeasures of acceleration industrialisation and urbanisation, logistics and transportation networks, and self-sufficiency development plan remain to be major challenges of this city cluster (ibid). The synergetic approach between multiple industries would help strengthen the region's not-yet-developed industrial chain. The region needs to focus more on spatial structures and distribution of load across the various districts, cities, and towns. In line with the guidelines of the new-type urbanisation plan (NUP), this approach could benefit from rural–urban development directions, achieving a more sustainable regional development plan, and meeting the carbon neutrality targets.

The recent proposal of developing a "Hydrogen Corridor" in November 2021 suggests a major regional shift in "*the first release of hydrogen fuel cell logistics vehicles*" [45]. This newly-launched bilateral industrial foundation is aimed to protect natural resources for "*the purpose of driving the construction of Chengdu-Chongqing economic circle as well as realising carbon peak and carbon neutrality*" (ibid). This new development is in line with the logistics development of the region, with two inter-city lines and one non-stop main line between the two major urban hubs of the region. The aim is also to link the urban hubs to other main urban regions of the cluster and link to other neighbouring clusters. As part of the industrial chain management plan, this new corridor will help link "*Zigong National Backbone Cold-chain Logistics Base, Southwest (Zigong) International Land Port, Jiulongpo Automobile Parts Distribution Base, and Lianglucuntan Bonded Port Area, forming an interconnected hydrogen-energy economy network with mutual integration*" (ibid). In doing so, a much stronger industrial linkage at the regional level is expected. This direction could boost the future development of hydrogen-energy industrials and fuel cell vehicle industries. By improving the upstream–downstream industrial chain, the new development is expected to be completed by 2025, leading to a later transition of expanding the hydrogen-energy industries. If to be completed successfully by 2030, the new infrastructural development could then support the next steps of meeting the carbon neutrality targets of 2060.

While the emphasis remains on developing two urban cores (i.e., the main urban areas of Chengdu and Chongqing), a joint action of dual carbon goal is proposed[46]. The two cities remain to have leading roles and core functions in the region, helping to address and strengthen the intersection between the two areas and the cities in between them. By catalysing the rural development, the aim is to find a sustainable approach towards driving rural development through the development of core urban areas (ibid). In doing so, the following directions are proposed in December 2021:

- *"To further integrate into the "domestic and international economic circulations", create an important growth pole and a new driving force for the nationwide high-quality development, and play a supporting role in promoting the development of China's western region in the new era, a leading role in the co-construction of the Belt and Road Initiative, and a demonstration role in promoting the green development of the Yangtze River Economic Belt;*
- *To strengthen all-round coordination and deepen the Sichuan-Chongqing strategic cooperation;*
- *To jointly forge a modern industrial system, strengthen government guidance, give play to the role of the market, promote the modernisation of the industrial chain and supply chain, foster industrial clusters with prominent competitive advantages, and build high-quality and distinctive industrial belts;*
- *To jointly build the Western (Chongqing) Science City, join hands in attracting investment and introducing talent, coordinate the construction of technology innovation platforms, and accelerate the construction of a technology innovation center with national influence;*
- *To jointly build the Western Financial Center, pool more financial resources, optimise financial products and services, and deepen financial opening-up and cooperation;*
- *To jointly promote the high-quality construction of the New International Land-Sea Trade Corridor (ILSTC), promote the layout of "logistics driven by corridors, trade driven by logistics, industries driven by trade", and facilitate the ILSTC to increase development speed, quality and benefits and reduce relevant costs.*
- *To go all out to achieve the goals of carbon peaking and carbon neutrality, pursue ecology-prioritised, green and low-carbon development, work together to reduce carbon emissions, control pollution, and expand green space, as well as accelerate the construction of a "green economic circle".*
- *To jointly provide public services, build a favourable environment for living, working and traveling, launch more Sichuan-Chongqing inter-regional government services, provide more high-quality public services, and improve the sense of gain of people and enterprises in the two places"* (ibid).

Hence, the dual carbon goal is still very much economic-oriented. However, to achieve the carbon peak targets of 2030 and carbon neutrality goals of 2060, the region is committed to increasing green development and low carbon transformation projects, helping accelerate the completion of ongoing major projects, like the national natural gas production base project (i.e., Shale Gas). Clean energy and synergies between industries remain to be the main target—and a calculable challenge—in an economic region of the energy mix and multiple industries.

References

1. CCTV Network (2021) The 10 provinces concentrated over 40% of the population, Chengdu-Chongqing area into a new flow. 这10省份集中全国四成人口,成渝地区成新流向. By CCTV International Network Co., Ltd. Available from: https://news.cctv.com/2021/05/14/ARTIzNw4KqQXezKAeVEpJ17i210514.shtml [3]
2. Chen C, LeGates R, Fang C (2019) From coordinated to integrated urban and rural development in China's megacity regions. J Urban Aff 41(2):150–169
3. Chen D, Zhou QG, Yu L (2020) Response of resources and environment carrying capacity under the evolution of land use structure in Chongqing section of the three Gorges reservoir area. J Environ Manag 274:111169
4. Cheshmehzangi A, Chen H (2021a) China's sustainability transitions: low carbon and climate-resilient plan for carbon neutral 2060. Springer Nature, Singapore
5. Cheshmehzangi A, Chen H (2021b) Key suggestions and steps ahead for China's carbon neutrality plan. In: China's sustainability transitions. Springer Nature, Singapore, pp 189–206
6. Cheshmehzangi A, Dawodu A, Sharifi A (2021) Sustainable urbanism in China. Routledge, New York and London
7. Cheshmehzangi A, Xie L, Tan-Mullins M (2021) Pioneering a Green Belt and Road Initiative (BRI) alignment between China and other members: mapping BRI's sustainability plan. Blue-Green Syst 3(1):49–61
8. Deng W, Cheshmehzangi A (2018) Eco-development in China: cities, communities and buildings. Palgrave Macmillan, Singapore
9. Fang C (2015) Important progress and future direction of studies on China's urban agglomerations. J Geog Sci 25(8):1003–1024
10. Feng L (2009) 2. New trends in China's regional economic development. In: Regional economic development in China. ISEAS Publishing, pp 9–27
11. Feng Y, Fei C (2015) Research on the integration of railway planning and urban planning. In: 2015 international conference on intelligent transportation, big data and smart city. IEEE, pp 669–672
12. Fu L, Zhai K (2018) A comparitive study of rural land property system reform in Chengdu and Chongqing. In: 3rd international conference on contemporary education, social sciences and humanities (ICCESSH 2018). Atlantis Press, pp 1324–1335
13. Gao H, Yang W, Wang J, Zheng X (2021) Analysis of the effectiveness of air pollution control policies based on historical evaluation and deep learning forecast: a case study of Chengdu-Chongqing region in China. Sustainability 13(1):206
14. Ge J (2021) Research on the impact of high-speed railway on the industrial structure change and carbon emission of Chengdu-Chongqing urban agglomeration. In: E3S web of conferences, vol 275. EDP Sciences, p 02001
15. Gong Q, Guo G, Li S, Liang X (2021) Decoupling of urban economic growth and water consumption in Chongqing and Chengdu from the "production-living-ecological" perspective. Sustain Cities Soc 75:103395
16. Gu D (2008) Some thoughts on building civilization of ecology: understanding of learning the report of 17th representative conference of the CPC. J Nantong Vocat Tech Shipp Coll 01. Available from: https://en.cnki.com.cn/Article_en/CJFDTotal-NTHY200801007.htm
17. iChongqing (2021) Chengdu, Chongqing to be a world-class city cluster. Available from: https://www.ichongqing.info/2021/01/14/chengdu-chongqing-to-be-a-world-class-city-cluster/
18. Han C, Zhang L (2017) The influence of university agglomeration on the high-tech industry development and employment growth. In: 2017 4th international conference on industrial economics system and industrial security engineering (IEIS). IEEE, pp 1–7
19. Hao Q-J (2016) Chengdu and Chongqing high-speed impact on regional cooperation of local government. Sch J Econ, Bus Manag 3(4):211–216. Available from: http://saspjournals.com/wp-content/uploads/2016/05/SJEBM-34211-216.pdf

20. He X, Yang J (2010) Chengdu-Chongqing economic zone: comparative advantage, underdevelopment and countermeasures. J Yibin Univ 2. Available from: https://en.cnki.com.cn/Article_en/CJFDTotal-YBSG201002020.htm
21. He X, Zhu X (2010) Research on industrial structuring and economic development of Chengdu-Chongqing economic zone. Soft Sci 6. Available from: https://en.cnki.com.cn/Article_en/CJFDTotal-XUXI201006017.htm
22. Hu J, Huang Y, Du J (2021) The impact of urban development intensity on ecological carrying capacity: a case study of ecologically fragile areas. Int J Environ Res Public Health 18(13):7094
23. Hu ZY, Xie YH, Chen ZQ, Wei J (2015) An analysis on evolution characteristics of spatial disparity of origin emissiveness in Chongqing. J Southwest China Norm Univ (Natural Science Edition) 12. Available from: https://en.cnki.com.cn/Article_en/CJFDTotal-XNZK201512012.htm
24. Huang X, Huang X, Liu M, Wang B, Zhao Y (2020) Spatial-temporal dynamics and driving forces of land development intensity in the western China from 2000 to 2015. Chin Geogra Sci 30(1):16–29
25. Lai HH (2002) China's western development program: its rationale, implementation, and prospects. Modern China 28(4):432–466
26. Li J, Qian Y, Zeng J, Yin F, Zhu L, Guang X (2020) Research on the influence of a high-speed railway on the spatial structure of the western urban agglomeration based on fractal theory—taking the Chengdu-Chongqing urban agglomeration as an example. Sustainability 12(18):7550
27. Li W (2009) Promote the Chengdu-Chongqing economic zone to set up eco-industrial system with the concept of circular economy. Soft Sci 4. Available from: https://en.cnki.com.cn/Article_en/CJFDTotal-XUXI200904019.htm
28. Li Y, Liu C, Zhang H, Gao X (2011) Evaluation on the human settlements environment suitability in the three gorges reservoir Area of Chongqing based on RS and GIS. J Geog Sci 21(2):346–358
29. Li Y, She C, Liu N, Zhang H, Zhang L, Zhu D (2016) Completion difficulties of HTHP and high-flowrate sour gas wells in the Longwangmiao Fm gas reservoir, Sichuan Basin, and corresponding countermeasures. Nat Gas Ind B 3(3):269–273
30. Liang X (2020) The evolution history of twin cities economic circle: an observation on the two cities in Chengdu-Chongqing region (2). 双城经济圈进化史——成渝地区双城经济圈系观察 (二). Special Chief of Sichuan Daily Observation. Available from: https://sichuan.scol.com.cn/ggxw/202007/57854021.html [2]
31. Liu H, Liu Y, Xu X (2012) Optimization of Industrial structure and layout in Chengdu-Chongqing economic zone based on environmental efficiency evaluation. Resour Environ Yangtze Basin 9. Available from: https://en.cnki.com.cn/Article_en/CJFDTotal-CJLY201209005.htm
32. Liu XY, Wang YQ (2008) Research into overall urban-rural development of Chengdu-Chongqing economic zone based on elliptic city group. J Chongqing Technol Bus Univ (West Forum) 2. Available from: https://en.cnki.com.cn/Article_en/CJFDTotal-CQSY200802004.htm
33. Luo R, Zhao J (2013) Analysis on the convergence of industrial structure in Chengdu-Chongqing economic zone and its policy options. AR Res Dev 05. Available from: https://en.cnki.com.cn/Article_en/CJFDTotal-DYYY201305007.htm
34. Ma Y, Johnson D, Wang JY, Shi X (2021) Competition for rail transport services in duopoly market: case study of China Railway (CR) express in Chengdu and Chongqing. Res Transp Bus Manag 38:100529
35. Mao M, Yu X (2013) Operation mechanism of Chengdu-Chongqing Logistics corridor. In: ICTE 2013: safety, speediness, intelligence, low-carbon, innovation, pp 1171–1177
36. National Development and Reform Commission (NDRC) (2016–2020) The 13th five-year plan for economic and social development of the People's Republic of China, p 219. Available from: https://en.ndrc.gov.cn/policies/202105/P020210527785800103339.pdf

37. National Development and Reform Commission (NDRC) and Land and Regional Research Group (2016) The strategic positioning and development goals of the Chengdu-Chongqing city cluster. 成渝城市群的战略定位和规划目标. Available from: http://theory.people.com.cn/n1/2016/0705/c83865-28526591.html
38. Pan Y, Ma L, Tang H, Wu Y, Yang Z (2021) Land use transitions under rapid urbanization in Chengdu-Chongqing region: a perspective of coupling water and land resources. Land 10(8):812
39. Qiu JQ, Zhu H (2004) Regional tourism combined development in the triangle zone of Sichuan, Guizhou, Chongqing. Geogr Geo-Inf Sci 2
40. Rao E, Xiao Y, Ouyang Z, Zheng H (2016) Changes in ecosystem service of soil conservation between 2000 and 2010 and its driving factors in southwestern China. Chin Geogra Sci 26(2):165–173
41. Shen A, Liu Y (2021) Countermeasures and suggestions for the co-construction and sharing of cultural information resources in Chengdu and Chongqing against the background of two-city economic circle. In: 2nd international conference on management, economy and law (ICMEL 2021). Atlantis Press, pp 224–230
42. Song J, Feng Q, Wang X, Fu H, Jiang W, Chen B (2019) Spatial association and effect evaluation of CO_2 emission in the Chengdu-Chongqing urban agglomeration: quantitative evidence from social network analysis. Sustainability 11(1):1
43. Science and Technology Daily (2021) Chengdu-Chongqing economic circle: a new growth driver in Western China. Available from: http://www.stdaily.com/English/WorldNews/2021-11/04/content_1229999.shtml
44. Tan Y, Yang Y, Leng B, LI, T. (2011) Urban network system in the Chengdu-Chongqing region in the perspective of advanced producer services. Prog Geogr 30(6):724–732
45. The People's Government of Sichuan Province (2021a) Sichuan, Chongqing Make joint effort in building "Hydrogen Corridor". Available from: https://www.sc.gov.cn/10462/10758/10760/10765/2021/12/3/907c24e8299b4bbd820e6d08a0720a9b.shtml
46. The People's Government of Sichuan Province (2021b) Fourth session of chongqing-sichuan party and government joint meeting for promoting construction of Chengdu-Chongqing economic circle kicks off. Available from: https://www.sc.gov.cn/10462/10758/10760/10765/2021/12/16/4c671af0c9c94777bd1d53f39489f181.shtml
47. Tian M, Tang N, Han J (2018) Research on the comprehensive economic strength of macro regional distribution in urbanization. In: Proceedings of the 21st international symposium on advancement of construction management and real estate. Springer, Singapore, pp 1225–1238
48. Wang C (2008) Ideas on industrial collaboration between Chengdu-Chongqing economic zone. J Chongqing Technol Bus Univ (West Forum) 2. Available from: https://en.cnki.com.cn/Article_en/CJFDTotal-CQSY200802003.htm
49. Wang F, Duan Y, Zhou J (2016) China's gas market under new situations: trends and countermeasures–taking Sichuan and Chongqing gas provinces as an example. Nat Gas Ind B 3(3):187–194
50. Wang K, Xiao L (2021) Grey theory based model applied in carbon emissions prediction for the Chengdu-Chongqing twin cities economic zone. In: 2021 international symposium on computer technology and information science (ISCTIS). IEEE, pp 146–149
51. Wang L (2019) Optimization of industrial structure in Chengdu-Chongqing economic zone. 成渝经济区产业结构优化浅析. Public Bus: Ventur Ed (11):2. Available from https://www.wenmi.com/article/px6rvb00a9of.html [4]
52. Wang R (2015) A critical review of regional economic integration in China. Turk Econ Rev 2(2):88–103
53. Wang S, Liu Y (2021) Research on the paths of Chengdu-Chongqing cultural cooperation from the perspective of cultural community. In: 2020 international conference on language, communication and culture studies (ICLCCS 2020). Atlantis Press, pp 193–197
54. Wang W, Xiong T, Zhang W, Luo B, Wang D, Jiang X, Rao Z, Jiang Y, Liu Y, Cheng H, Luo X. (2020). Observation and analysis of VOCs in nine prefecture-level cities of Sichuan Province, China. Environ Monit Assess 192(8):1–14

55. Wang X, Gao X, He R (2018) Research on the investment cooperation between Sichuan and Chongqing based on the motives of manufacturing transformation and upgrading. In: Proceedings of the 4th international symposium on social science (ISSS 2018). Dalian, China, pp 12–13
56. Wang YC, Ma ZL (2005) A cluster analysis about the economic level of West-Chongqing economic hallway. J Chongqing Inst Technol 4. Available from: https://en.cnki.com.cn/Article_en/CJFDTotal-CGGL200504003.htm
57. Wei Q, Tang Y, Ma Y (2021) A comparative study on the selection and cultivation of pillar industries in Chengdu–Chongqing region of China. In: The 2nd international conference on economics, education and social research (ICEESR 2021). Available from: https://www.webofproceedings.org/proceedings_series/ESSP/ICEESR%202021/ZJ_DAM19839.pdf
58. Wu H, Li Z, Ke G (2016) Characteristics of the Changxing Fm biohermal gas reservoir in the Yuanba Gasfield, Sichuan Basin and development countermeasures. Nat Gas Ind B 3(6):537–544
59. Wu X, Wu P, Guo Z (2021) The correlations of transportation with regional economy and demographic space. J Intell Fuzzy Syst (Preprint) 1–11. Available from: https://content.iospress.com/articles/journal-of-intelligent-and-fuzzy-systems/ifs189967
60. Xie H, Wang W (2015) Spatiotemporal differences and convergence of urban industrial land use efficiency for China's major economic zones. J Geog Sci 25(10):1183–1198
61. Xing X, Qi S, Zhang J, Wu C, Zhang Y, Yang D, Odhiambo JO (2011) Spatial distribution and source diagnosis of polycyclic aromatic hydrocarbons in soils from Chengdu economic region, Sichuan Province, Western China. J Geochem Explor 110(2):146–154
62. Xiong SX, Fang XB (2010) the impact of complementary assets on enterprise's technological innovations: a empirical study based on chinese listed companies. Econ Manag J 6
63. Xu C (2010) The origin of the Chengdu-Chongqing economic zone and its practical problems. 改革 3. Available from: https://en.cnki.com.cn/Article_en/CJFDTotal-REFO201003011.htm
64. Xu X (2014) Industrial cluster and logistics network to promote the new development of the Chengdu rail transit industry. In: ICLEM 2014: system planning, supply chain management, and safety, pp 1296–1300
65. Yang C, Han H, Song J (2014) Spatial distribution of migration and economic development: a case study of Sichuan Province, China. Sustainability 6(10):6509–6528
66. Yang C, Zeng W, Yang X (2020) Coupling coordination evaluation and sustainable development pattern of geo-ecological environment and urbanization in Chongqing municipality, China. Sustain Cities Soc 61:102271
67. Yang J, Huang X, Tian J (2013) Building ecological civilization: some thinking and countermeasures. Comp Econ Soc Syst 03. Available from: https://en.cnki.com.cn/Article_en/CJFDTotal-JJSH201303014.htm
68. Yao Z, Ye K, Xiao L, Wang X (2021) Radiation effect of urban agglomeration's transportation network: evidence from Chengdu–Chongqing urban agglomeration. China. Land 10(5):520
69. Zhang C, Yang J (2018) Analysis and forecast of coupling coordination development among the ESRE system: a case study of Chongqing. JOurnal Chongqing Univ Technol (Social Science) 09
70. Zhang Q, Hu Y, Liu J, Liu Y, Ren W, Li J (2011) Identification of urban clusters in China based on assessment of transportation accessibility and socio-economic indicators. Acta Geogr Sin 66(6):761–770
71. Zheng Y, Yue B, Han X, Li Z (2021) Research on the impact of industrial development on energy utilization efficiency under the background of urban agglomeration planning. In: E3S web of conferences, vol 283. EDP Sciences
72. Zhou C, Fu H (2021) An analysis of urban art ecology under the background of Chengdu-Chongqing double cities economic circle. In: 2nd international conference on language, art and cultural exchange (ICLACE 2021). Atlantis Press, pp 518–521

73. Zhou W, Zhang B, Zhang B (2010) Spatial relationship between dual-core cities based on gravity model—a case of Chengdu and Chongqing City. Resour Dev Mark 26(11):82–85
74. Zhu M, Shao L (2020) An analysis on the economic cooperation and the industrial synergy of the main river region: from the perspective of the Yangtze River economic zone. J Ambient Intell Humaniz Comput 11(3):1055–1064

Chapter 7
Yangtze Mid-River Delta: China's Strategic Fulcrum of Dynamic Growth Pole in the Form of Middle Triangle

7.1 General Introduction to the Yangtze Mid-River Delta City Cluster or Middle Triangle

The middle reaches of the Yangtze River, also known as the "Middle Triangle," is a large national city cluster with Wuhan as the center, Wuhan Metropolitan Area, Changsha-Zhuzhou-Xiangtan City Area, and Poyang Lake Area as the main body. It is commonly known as the Yangtze Mid-River Delta. The cluster is in the form of a triangle between the main cities of the provinces included in this large inland cluster. The planning scope includes cities and regions of Wuhan, Huangshi, Ezhou, Huang Gang, Xiaogan, Xianning, Xian Tao, Qianjiang, Tianmen, Xiangyang, Yichang, Jingzhou, and Jingmen in Hubei Province. It also includes Hunan's capital city of Changsha and other cities and regions of Zhuzhou, Xiangtan, Yueyang, Yiyang City, Changde, Hengyang, and Loudi Cities. It also includes cities in Jiangxi province, including Nanchang, Jiujiang, Jingdezhen, Yingtan, Xinyu, Yichun, Pingxiang, Shangrao, Fuzhou, and Ji'an city part counties (or districts). This large city cluster covers a total area of 326.1 thousand square kilometers (in 2017), with an estimated population of 175 million (in 2019), and an overall GDP of 7.9 trillion Yuan [25].

As a new hub for steel stocks [43], the Yangze Mid-River Delta (YMRD) is a significant hub in the context of China. While it is classified as a middle city cluster, its large size and significance are at the level of Chinese higher-level city clusters. Studies on the impact of landscape patterns on ecosystem services in this region, such as the one by [], suggest the importance of environmental factors that should be considered in future sustainable development plans. As planned by the local and national level governmental units, the middle reaches of the Yangtze River or YMRD city cluster is expected to become China's core growth pole [14, 39]. The main trend of urbanisation in the last 50 years is replacing mega-cities by city clusters. Similar examples to YMRC exist elsewhere, where a sort of triangle is formed between major urban hubs. An example of a major urban cluster in Europe is the formation of a cluster between London, Manchester, and Northwest Europe. There is also a similar

A. Cheshmehzangi and T. Tang, *China's City Cluster Development in the Race to Carbon Neutrality*, Urban Sustainability, https://doi.org/10.1007/978-981-19-7673-5_7

major city cluster in North America, such as the Great Lakes. In fact, in the USA alone, there are at least 11 city cluster plans as mega-regions, which are expected to be accomplished by 2050 [18]. The model of such city cluster development (CCD) is related to mega-urban governance approaches that are becoming more visible in larger countries or sub-regions. East Asia also has two large city clusters, namely, Japan's Tokaido city cluster and China's Yangtze River Delta city cluster. However, the city cluster in the middle reaches of the Yangtze River is expected to become the core growth pole [17, 70] in the next stage of China's development. The two factors of industrial green spatial pattern evolution and the creation of an economic belt along the Yangtze River [20, 22, 29] are the backbones of making such a growth pole in the inland regions. As part of China's national-level economic development planning, the YMRD city cluster could become a major economic hub and role model to other middle-level and small city clusters.

Urban agglomeration in the middle reaches of the Yangtze River has become a vital fulcrum in central China. This cluster is known as a national strategic fulcrum [35], mostly known for economic and industrial development in the middle-inland regions of China. Enhancing the water resources environment and high-quality economic development [74] are among the few factors aligned with China's newly-development high-quality urbanisation plan [9, 11, 12]. Compared with other city clusters, the city cluster's biggest feature in the Yangtze River's middle reaches is the word "Zhong" (meaning 'middle'). As a middle region connecting the east and the west, connecting the south and the north, the city cluster in the middle reaches of the Yangtze River does have some conundrums. One end is the eastern coastal areas, which has been the first breakthrough in the Yangtze River Delta, Beijing-Tianjin-Hebei, and the Pearl River Delta market competition advantage. At the other end, the cluster includes the underdeveloped western region, which is the priority target of financial support.

As an important fulcrum in the middle of the "balance," the middle reaches of the Yangtze River city cluster also has their own unique advantages [39]. From the area, the population scale, and the structure of city clusters, the potential and space of city cluster in the middle reaches of the Yangtze River are better than those in the Changjiang Delta and the Zhujiang Delta [13, 30, 66]. The development of city cluster in the middle reaches of the Yangtze River is expected to bring many benefits. At the national level, the YMRC is expected to expand the new space of regional development. The overall implementation of the strategic combination of four plates and three supporting zones is the backbone of infrastructural and industrial development in the regions covered under this city cluster. More importantly, the middle reaches of the Yangtze River city cluster is an obvious, comprehensive transportation hub network and has a vital fulcrum role. Its significance in terms of transportation and connectivity is paramount nationally and internationally. It was also where the COVID-19 disease was spread within days and became a major pandemic in only a couple of months. At the same time, we can argue that the development of the city cluster in the middle reaches of the Yangtze River plays an important role in the completion of the task of "*guiding 100 million people to the nearest urbanization in the central and western regions*" [6, 38, 41]. Thus, the YMRD city cluster is more of a role model requiring higher-level attention. When we attended the EU-China urbanisation forum in Beijing, Premier Li Keqiang attended the event and highlighted the importance of economic growth in regions such as the YMRD city cluster. It was from then that the significance of such clusters became more visible with further national-level recognition and support.

7.2 A Micro-Historical Analysis of the Yangtze Mid-River Delta City Cluster

The three provinces of Hunan, Hubei, and Jiangxi have maintained close economic and trade contacts and cultural exchanges since ancient times [12, 16, 23, 42, 63]. After the founding of the People's Republic of China (PRC), especially since the reform and opening up in 1978, the local governments and people of the three provinces have increasingly close contacts, which laid the foundation for the construction of the city cluster in the middle reaches of Yangtze River in Hunan, Hubei and Jiangxi Provinces [39]. The middle reaches of the Yangtze River have been the granary of the whole country since ancient times, with a solid agricultural production base [41]. It is one of the areas with the most abundant agricultural products and the most active agrarian product trade in China [45, 68]. This area became one of the cradles of China's modern industrial revolution after Zhang Zhidong became governor of Huguang in 1889 [4, 33, 50]. The middle reaches of the Yangtze River is the birthplace of China's modern industrial revolution. Later, under the planned economic system, it became an old industrial base supported by the state for a long time [64]. Before and after the reform and opening up, the three provinces of Hunan, Hubei, and Jiangxi started extensive cooperation [50, 65]. Under the developing central China vision [27], a new regional development was started for manufacturing and industrial development.

After entering the new century, non-governmental cooperation is more in-depth and extensive [39]. After 2012, the explosive growth of local and non-governmental trade between the three provinces promoted the formation of the city cluster in the middle reaches of the Yangtze River. In terms of its population structure, the region has some deficiencies. In 2019, the population in the middle reaches of the Yangtze River showed negative growth in the indicator of new migrants. In 2019, the city cluster in the middle reaches of the Yangtze River had a net outflow of 250,000 people, most of whom will flow to the more developed Yangtze and Pearl River Delta regions. The trend has continued since then and is expected to become a bottleneck for future innovation and industrial investment at the regional level.

7.3 An Overview of Industrial Structure in the Yangtze Mid-River Delta

The city cluster in the middle reaches of the Yangtze River is a multi-center city cluster consisting of Wuhan City Circle, Changsha-Zhuzhou-Xiangtancity cluster, Poyang Lake Ecological Economic Zone, and Wanjiang City. The cluster has excellent potential to become the fourth economic growth pole after the Yangtze River Delta, the Pearl Delta, and the Bohai Rim [41]. However, it is still in the initial stage of development. The development space, industrial foundation, human resources, ecological environment, and radiation domestic market have obvious competitive advantages.

As per the governmental reports, the development potential of the YMRD is huge, but the overall strength needs to be improved. The robustness of central cities has increased but still needs to be improved [39]. Division of labour and cooperation has begun to take a foundation to be deepened, which would need at least another decade to show some of the results. The cities in the middle reaches of the Yangtze River have begun to form the basis of division of labour and cooperation [40, 46]. There is an emphasis on industrial synergy and economic cooperation as two primary focuses in the region [74]. This is similar to the idea of shifting 'from metropolis to polyoposis', as [19] suggested, representing the region's development in the past two decades.

Furthermore, the urban spatial connection of the city cluster in the middle reaches of the Yangtze River presents three main characteristics: First, the central cities have strong radiation ability and are closely connected with each other. Secondly, the internal gravitation of the city cluster in four provinces is remarkable. And third, the economic links among the four provincial city clusters need to be improved [14]. The convergence of industrial structure is essential in weakening the attraction of labor division and cooperation among the city cluster. The industrial structure emphasises heavy industry, such as automobiles, metallurgy, coal, etc. However, the development of new service industries is lagging behind. The agricultural development of city clusters in the four provinces all stresses extensive traditional agriculture [3]. Additionally, the following two factors need further attention:

(1) ***Coordination mechanisms are not yet mature***
Compared with the Yangtze River Delta city cluster, the Delta is a government, industry, enterprise interaction, and gradually clear transportation, energy, environmental protection, credit, social security, and other critical areas of cooperation. While the city cluster in the middle reaches of the Yangtze River is mainly guided and promoted by the Government, the spontaneous cooperation behavior of enterprises and society is still insufficient. Thus, the cooperation mechanism has become mature [39].

(2) ***Ecological protection faces severe challenges***
Affected by natural and human factors, the ecological problems of the city cluster in the middle reaches of the Yangtze River are increasingly severe. Both Dongting Lake and Poyang Lake have severe soil erosion and desertification

[21, 44, 53, 56, 57]. In addition, industrial pollution, agricultural pollution, and air pollution are also very serious. These pollution situations have led to lake eutrophication, which has become increasingly prominent in recent years. With the title of "Province of Thousand Lake" [8], the number and area of lakes in Hubei Province have been reduced rapidly due to natural siltation and artificial reclamation over the years.

7.4 Overview of the Sustainability Agenda for the Yangtze Mid-River Delta City Cluster: Plans and Perspectives

The YMRD city cluster aims to pursue sustainable development, build an ecological civilization, and establish and improve cross-regional mechanisms for building ecological civilization [60, 67]. The governmental plan is to formulate and implement master plans for the environment of cities in the cluster, promote ecological integration in strict accordance with the orientation of the main functions, strengthen comprehensive management of the ecological environment, promote green development of the city cluster, and form a pattern of harmonious development between man and nature. Below, we summarised the plan and perspectives in three perspectives of environmental, economic, and social.

7.4.1 Sustainable Development Plan from the Environmental Perspective

The preliminary plan is to build ecological barriers together and in between the cities of the region. In doing so, the aim is to jointly protect water resources and the water environment of the city cluster and region. The plan is to strengthen ecological water protection and environmental water management of lakes and wetlands [55]. The government proposes implementing red lines for water resource development and utilisation, water use efficiency, and strictly controlling the total amount of pollutants discharged. They aim to focus on protecting drinking water sources and optimising the distribution of industries in the Yangtze River's mainstream to promote ecological restoration [39]. Hence, this approach could strengthen the construction of state-level aquatic germplasm resources reserves, wetland nature reserves, and wetland parks, strengthen wetland ecological restoration, and restore the wetland's functions of purifying water quality, adjusting climate, and maintaining biodiversity [7]. Such major plans require support. Thus, the central government plans to support the construction of pilot cities to promote a water-ecological civilization, and strengthen the comprehensive improvement of rural river courses [39]. Such investment would strengthen the improvement of river outfalls, the construction of urban and rural sewage and garbage treatment facilities, and improve sewage collection networks and transportation systems. The ongoing projects and initiatives target accelerating the construction of centralised sewage treatment plants in industrial parks [24, 36, 58]. To strengthen the supervision and management of sewage outlets into the river, the primary target is to develop rational optimisation and adjustment of the layout of sewage outfalls into rivers. The government then aims to increase efforts to reduce agricultural non-point source pollution, delimit areas where livestock and poultry are banned or restricted, and build waste treatment and storage facilities on livestock and livestock farms.

The other ongoing initiative is building the city cluster's "green heart" [26]. Construction of the city cluster ecological barriers with Mufu Mountain and Luoxiao Mountain [5] as the main body and green corridors along rivers, lakes, and major traffic axes as the link is the backbone of this overarching plan—i.e., building city cluster "green heart." The ongoing programmes close hillsides for afforestation to strengthen comprehensive control of soil erosion, strictly implement the system of water and soil conservation plans for production and construction projects following the law and strengthen supervision and management of soil and water conservation in the development and construction of all types of projects to prevent new man-made soil erosion. The local and regional governments also aim to promote the construction of public ecological forests, improve the structure of forests, and strictly control logging and mining. They suggest strengthening the development of nature reserves, scenic spots, forest parks, and geoparks, strengthening the protection of biodiversity, and building an ecological network with excellent ecology, perfect functions, and beautiful landscapes [15, 73]. To consider the "green heart" involved in the relevant counties (i.e., cities and districts) into the scope of national critical ecological function areas.

Lastly, in recent years, the focus has been on constructing ecological corridors. Through this initiative, the government aims to strengthen the construction of green belts along transportation routes and on both banks of rivers and speed up the development of key projects such as the shelterbelt system along the Yangtze River, the Three Gorges Reservoir Area, the protective forest belt around Poyang and Dongting Lakes, the demonstration of carbon sequestration, and green belt along expressways, and railways. To achieve the key targets, efforts should be made to improve the ecological quality of the Dabie Mountains, Dahongshan, Dameishan, Huaiyushan, Luoxiao Mountain, Hengshan Mount, Lushan Hill, Wugong Mount, and the northern section of Wuling Mountain [41]. The projects plan to jointly build the Yangtze River system, wetlands, mountains, road green belt, and farmland forest network as the network ecological corridor's main framework and enhance the ecosystem's function. The government plans to strengthen the improvement of sloping cultivated land and the matching of slope water system projects in hilly areas, control soil erosion under forests, and control the collapse of hills. The initiative promotes the construction of clean small river basins in the surrounding areas of cities and further strengthens key projects for soil and water conservation in old revolutionary base areas.

7.4.2 Sustainable Development Plan from the Economic Perspective

The plan for developing a growth pole is the most prominent target in the region. This plan is boosted by the vigorous development of a circular economy. In this regard, the ongoing projects are to build an industrial chain of resource recycling between enterprises and parks, using development zones and industrial parks that meet the requirements[39, 41, 54, 58]. The government proposes comprehensively promoting industrial parks' recycling transformation and encouraging national development zones to establish national circular economy demonstration zones and national eco-industrial demonstration zones [2, 61]. The plan is also to strengthen the clean production audit and do a good job in iron and steel, petrochemical, nonferrous metals, machinery manufacturing, building materials, and papermaking industries clean production [11]. It also strengthens the comprehensive use of resources, vigorously promotes the comprehensive utilisation of solid industrial wastes such as phosphogypsum and smelting slag [52], and focuses on building demonstration bases for the utilisation of renewable resources []. Aligned with the project plans, the government supports the development of the remanufacturing industry, builds a recycling system for renewable resources, promotes recycling and recycling of renewable resources, and takes the lead in establishing a comprehensive recycling network for urban clusters along the middle reaches of the Yangtze River [39]. As the backbone of implementing the strategies, the government supports cities with conditions to actively carry out the construction of "urban minerals", food waste utilization and harmless treatment, renewable resources processing, and utilisation [39]. It also actively builds an agricultural circular economy industrial chain [] and promotes agricultural and forestry waste recycling.

7.4.3 Sustainable Development Plan from the Social Perspective

The main plan is to advocate a green and low-carbon lifestyle. The government gives full play to the role of the Hubei National Low-carbon Province Pilot Program, Jiangxi National Ecological Civilization Demonstration Zone, Wuhan, Nanchang, and Jingdezhen National Low-Carbon City Pilot, and the State Sustainable Development Experimental Zone [10, 30, 39]. The engaged stakeholders through the relevant governmental units and departments support the creation of low-carbon enterprises, low-carbon parks, low-carbon communities, and low-carbon cities. They vigorously promote green buildings, energy-saving and land-saving residences, fully decorated residential units, energy-saving electrical appliances, and water-saving equipment adapted to hot summer and cold winter climate zones [62]. The intention is to gradually promote the separation of household waste recycling, reduce the use of disposable products, and limit excessive packaging [39]. Other plans are advocating green travel, encouraging consumers to buy energy-saving environmental protection vehicles, and vigorously developing public transport and a slow urban system. These projects implement green label certification systems such as organic product certification and good agricultural practice certification. As has occurred in recent years, the government gives priority to environmental label products and energy-saving products.

7.5 Pathways to Achieve China's Carbon Neutral Plan: Opportunities and Challenges

The first pathway is to jointly build a cross-regional environmental protection mechanism. In doing so, the YMRD city cluster could strengthen the joint prevention and treatment of environmental pollution [72]. Through strengthening the cooperation in environmental access and management, the government aims to gradually unify the standards for environmental access for industrial projects and construction projects in the city cluster. It also seeks to set standards for the discharge of major pollutants, advance the building of a credit system for environmental protection, and explore the establishment of a "blacklist" system for protection of environmental protection. In doing so, it is intended to accelerate the establishment of mechanisms for encouraging environmental compliance and punishing dishonesty in this particular area. In addition, the government aims to strengthen cooperation in emergency response mechanisms, establish a rapid notification mechanism for environmental emergencies, and jointly address unexpected regional ecological and environmental issues. To date, they have jointly implemented the Strategic Action Plan for Water Environmental Protection [48], accelerated the establishment of the Internet of Things (IoT) for joint prevention and control of water pollution, strengthened the assessment and management of cross-border water quality sections and key sections, and promote the prevention and treatment of transboundary water pollution.

One of the critical aspects of delivering a carbon neutrality plan for the region is to focus on pollution reduction. So far, the local governments have worked together to prevent and control air pollution, implement the Clean Air Action Plan for Cities (CAAPC) [69], and comprehensively strengthen the prevention and control of air pollution in key regions and industries [14, 41]. This approach helps to strengthen coordinated control over the emission of industrial dust, dust, urban dust, and volatile organic compounds. It also helps to vigorously promote the construction of projects to remove sulfur and denitrification, strengthen the elimination of yellow-standard vehicles and old vehicles, and treat motor vehicle pollution [39]. The other plans are to reduce contaminated areas like the Zhuzhou Qingshuitang [28] and other urban old industrial regions. These are either completed or ongoing through relocation and transformation efforts. The local governments also strengthen the joint prevention and control of heavy metal pollution, establish an environmental early warning system for key regions and key river basins, and carry out comprehensive regional improvement with critical regions as the core and the Xiangjiang River Basin [1, 47] as the focus.

The second overarching pathway is to improve the ecological compensation mechanism at both the provincial and regional levels [71]. In accordance with the principle that the developer shall protect and the beneficiary shall compensate, various ecological compensation methods are explored in the areas of forests, wetlands, river basin water resources, and mineral resources [39]. Accordingly, the government promotes the establishment of horizontal ecological compensation mechanisms between downstream and upstream areas, development and protection areas, and ecological beneficiary areas and ecological protection areas. They aim to establish and improve ecological compensation systems for drinking water sources, nature reserves, key ecological functional areas, mineral resources development, and river basin water environment protection. They continue to carry out the pilot project of national ecological compensation at the source of the Fuhe River and including Poyang Lake, Dongting Lake, and other important wetlands in Hunan, Yuanli, Sishui, Honghu, and the middle and lower reaches of the Hanjiang River [31, 37, 39, 60]. The government also supports the establishment of the South China Forest Carbon Sink Fund, promotes carbon sequestration afforestation and carbon emission reduction targets paid for the use of transactions and supports the construction of regional forest rights trading markets by the Southern Forestry Property Exchange and the Central Forestry [14, 39]. The continuous efforts also support the construction of the Hubei Carbon Emission Rights Trading Center [32] and encourage Xinyu and other regional carbon emission rights trading markets.

Lastly, the plan is also to implement the environmental regulatory law enforcement linkage. This implementation plan includes unified law enforcement standards, strict implementation of major pollutants total control, environmental impact assessment, construction project environmental protection facilities "three simultaneous," deadline treatment, regional river basin industry limited approval, listing supervision, and environmental protection after the inspection system. Therefore, the focus is given to environmental protection strategies to reverse some of the negative trends of earlier rapid development in the region, particularly the infrastructural and environmental developments. The government aims to establish and improve cross-administrative tracking mechanisms, consultation mechanisms, and arbitration mechanisms for environmental governance. In doing so, the ongoing initiatives strengthen joint supervision and dispute mediation. As part of the environmental protection plans, joint defense mechanisms are set against alien pests' invasion to protect national biosafety [39, 49]. Other efforts strengthen energy conservation supervision and environmental protection law enforcement team building. The longer-term plan is also to improve the intensity of environmental law enforcement in line with carbon intensity and carbon peak plans. The starting point is for the environmental protection departments to form unified supervision and allocate tasks to the relevant departments who are responsible for the environment law enforcement mechanism. As environmental factors are the main factors for consideration in the YMRD city cluster, the plans are in place to ensure environmental protection strategies could benefit the foreseeing carbon neutrality pathways of the region.

References

1. An Q, Wu Q, Li J, Xiong B, Chen X (2019) Environmental efficiency evaluation for Xiangjiang River basin cities based on an improved SBM model and Global Malmquist index. Energy Econ 81:95–103
2. Bai X, Chu CL, Ju MT (2012) Research on construction of national eco-industrial demonstration park of China. In: Applied mechanics and materials, vol 178, Trans Tech Publications Ltd., Bäch, pp 825–828
3. Boccaletti G (2021) Water: A Biography. Pantheon Books, New York City
4. Chang A (2017) Reappraising Zhang Zhidong: forgotten continuities during China's self-strengthening, 1884–1901. J Chin Mil Hist 6(2):157–192
5. Chen D, Pan Y, Jin X, Du H, Li M, Jiang P (2021) The delineation of ecological redline area for catchment sustainable management from the perspective of ecosystem services and social needs: A case study of the Xiangjiang watershed, China. Ecol Indic 121:107130
6. Chen M, Gong Y, Li Y, Lu D, Zhang H (2016) Population distribution and urbanization on both sides of the Hu Huanyong line: answering the Premier's question. J Geog Sci 26(11):1593–1610
7. Chen W (2020) The transformation of agricultural production system in line with "Resource and Ecologically Sound Agriculture". In: Challenges and opportunities for Chinese agriculture. Palgrave Macmillan, Singapore, pp 183–340
8. Chen X, Hu T, Ren F, Chen D, Li L, Gao N (2014) Landscape analysis of geographical names in Hubei province, China. Entropy 16(12):6313–6337
9. Cheshmehzangi A (2016) China's new-type Urbanisation Plan (NUP) and the foreseeing challenges for decarbonization of cities: a review. Energy Procedia 104:146–152
10. Cheshmehzangi A (2020) Towards urban energy management: a brief review of China's recent transition in urban energy demand. Low Carbon Econ 11(01):1
11. Cheshmehzangi A (2021) Low carbon transition at the township level: feasibility study of environmental pollutants and sustainable energy planning. Int J Sustain Energ 40(7):670–696
12. Cheshmehzangi A, Dawodu A, Sharifi A (2021) Sustainable urbanism in China. Routledge, New York
13. China Services Info (2016) Mid-Yangtze river city cluster seeks further integration, Available from: https://govt.chinadaily.com.cn/s/201603/03/WS5d131d6b498e12256565eaad/mid-yangtze-river-city-cluster-seeks-further-integration.html
14. Chinese Government (2015) Urban agglomeration development planning in the middle reaches of the Yangtze River 长江中游城市群发展规划 Available from: http://www.gov.cn/xinwen/site1/20150416/747814291881420S3.pdf
15. Deng B, Affolderbach J, Deutz P (2020) Industrial restructuring through eco-transformation: green industrial transfer in Changsha–Zhuzhou–Xiangtan, Hunan Province. Sustainability 12(17):6945
16. Dillon M (2008) Contemporary China-an introduction. Routledge, New York
17. Ge F, Chen W, Zeng Y, Li J (2021) The nexus between urbanization and traffic accessibility in the middle reaches of the Yangtze River urban agglomerations, China. Int J Environ Res Public Health 18(7):3828
18. Groff SP, Stefan R (2019) China's city clusters: pioneering future mega-urban governance. Am Aff 3(2). Available from: https://americanaffairsjournal.org/2019/05/chinas-city-clusters-pioneering-future-mega-urban-governance/
19. Hall P, Pain K (2012) From metropolis to polyopolis. The polycentric metropolis. Routledge, New York, pp 18–31
20. He J, Li C, Yu Y, Liu Y, Huang J (2017) Measuring urban spatial interaction in wuhan urban agglomeration, central China: a spatially explicit approach. Sustain Cities Soc 32:569–583
21. Hu W, Li G, Li Z (2021) Spatial and temporal evolution characteristics of the water conservation function and its driving factors in regional lake wetlands—two types of homogeneous lakes as examples. Ecol Ind 130:108069

22. Huang S, Huang Y, Zhang X, Sheng S, Mao L, Huang X, Liu R, Huang B (2021a) Analysis of the spatio-temporal evolution and factors influencing the freight network in the middle reaches of the Yangtze River. In: Discrete dynamics in nature and society
23. Huang W, Xi M, Lu S, Taghizadeh-Hesary F (2021) Rise and fall of the grand canal in the ancient Kaifeng city of China: role of the grand canal and water supply in urban and regional development. Water 13(14):1932
24. Huang, Y., & Wang, J. (2015). Research on Establishment of an Index System for the Low-carbon Industrial Parks in Wuhan City. In (2015) international symposium on material, energy and environment engineering. Atlantis Press, Amsterdam, pp 329–333
25. Intermediate City Planning and Design Research Institute (2020) Center for city cluster in the Middle Reaches of the Yangtze River 长江中游城市群研究中心. Available from: http://www.reportway.org/changjiangzhongyou.html
26. Jiao S, Gao Q, Wei CY, Peng SN, Liu S (2012) A study of the protection-oriented development strategy of green heart in hilly areas: the case of green heart demonstration zone of Changsha-Zhuzhou-Xiangtan Agglomeration. In: Advanced materials research, vol 433. Trans Tech Publications Ltd., pp 1280–1285
27. Lai H (2007) Developing central China: a new regional programme. China: Int J 5(1):109–128
28. Lei M, Zeng M, Wang LH, Williams PN, Sun GX (2010) Arsenic, lead, and cadmium pollution in rice from Hunan markets and contaminated areas and their health risk assessment. Acta Sci Circumst 30(11):2314–2320
29. Li L, Liu Y (2017) Industrial green spatial pattern evolution of Yangtze river economic belt in China. Chin Geogra Sci 27(4):660–672
30. Li S, Xu Q, Liu J, Shen L, Chen J (2022) Experience learning from low-carbon pilot provinces in China: pathways towards carbon neutrality. Energ Strat Rev 42:100888
31. Li X, Yu X, Jiang L, Li W, Liu Y, Hou X (2014) How important are the wetlands in the middle-lower Yangtze River region: an ecosystem service valuation approach. Ecosyst Serv 10:54–60
32. Liu L, Luo J, Xiao X, Hu B, Qi S, Lin H, Zu X (2022) Spatio-temporal evolution of urban innovation networks: a case study of the urban agglomeration in the middle reaches of the Yangtze River, China. Land 11(5):597
33. Liu Y (2010) Revisiting Hanyeping Company (1889–1908): a case study of China's early industrialisation and corporate history. Bus Hist 52(1):62–73
34. Liu Y, Tian L, Sun H, Yuan L, Zhang X (2022) Marginal return-ability measurement of carbon emission right and its application to unification route analysis of carbon markets. J Clean Prod 345:130684
35. Lu S, Guan X, He C, Zhang J (2014) Spatio-temporal patterns and policy implications of urban land expansion in metropolitan areas: a case study of Wuhan urban agglomeration, central China. Sustainability 6(8):4723–4748
36. Lv J, Hou J (2011) Analysis of rural ecological environment governance in the two-oriented society construction: a case study of Xiantao City in Hubei Province. Procedia Environ Sci 11:1278–1284
37. Mei X, Dai Z, Fagherazzi S, Chen J (2016) Dramatic variations in emergent wetland area in China's largest freshwater lake, Poyang Lake. Adv Water Resour 96:1–10
38. Ministry of Foreign Affairs of the People's Republic of China (2015) Li Keqiang attends and addresses China-EU urbanization partnership forum 2015. Available from: https://www.fmprc.gov.cn/mfa_eng/gjhdq_665435/3265_665445/oumeng_eng_664510/Activitiesoumeng_664514/201507/t20150702_573840.html
39. National Development and Reform Commission (NDRC) (2015) Urban Agglomeration development planning in the middle reaches of the Yangtze River 国家发展改革委关于印发: 长江中游城市群发展规划的通知 Available from: https://www.ndrc.gov.cn/xxgk/zcfb/tz/201504/t20150416_963800.html?code=&state=123
40. Niu F, Yang X, Wang F (2020) Urban agglomeration formation and its spatiotemporal expansion process in China: from the perspective of industrial evolution. Chin Geogra Sci 30(3):532–543

41. Qin Z, Zhang J (2013) Deep historical origin of constructing city cluster in the middle reaches of the Yangtze River. 构建长江中游城市群的深厚历史渊源. Douding Network https://www.docin.com/p-669601729.html
42. Shi X, Heerink N, Qu F (2007) Choices between different off-farm employment sub-categories: an empirical analysis for Jiangxi Province, China. China Econ Rev 18(4):438–455
43. Song L, Dai S, Cao Z, Liu Y, Chen WQ (2021) High spatial resolution mapping of steel resources accumulated above ground in mainland China: past trends and future prospects. J Clean Prod 297:126482
44. Song W, Liu M (2017) Farmland conversion decreases regional and national land quality in China. Land Degrad Dev 28(2):459–471
45. Sun ZL, Li X (2018) The trade margins of Chinese agricultural exports to ASEAN and their determinants. J Integr Agric 17(10):2356–2367
46. Tang C, Fan J, Sun W (2015) Distribution characteristics and policy implications of territorial development suitability of the Yangtze River Basin. J Geog Sci 25(11):1377–1392
47. Tian Y, Xu YP, Wang G (2018) Agricultural drought prediction using climate indices based on support vector regression in Xiangjiang River basin. Sci Total Environ 622:710–720
48. Tian Z, Tian Y, Chen Y, Shao S (2020) The economic consequences of environmental regulation in China: from a perspective of the environmental protection admonishing talk policy. Bus Strateg Environ 29(4):1723–1733
49. Wan FH, Yang NW (2016) Invasion and management of agricultural alien insects in China. Annu Rev Entomol 61(1):77–98
50. Wang B (2018a) Heinrich Hildebrand: a German railway engineer in late Qing China. Chin Ann Hist Sci Technol 2(1):34
51. Wang L (2018b) Emerging urban-regions in central China: the case of the upper Yangtze River Delta. In: Urbanization and urban governance in China. Palgrave Macmillan, New York, pp 49–71
52. Wang JA, Liang S, Shi P (2022b) Environmental pollution and protection. In: The geography of contemporary China. Springer, Cham, pp 215–229
53. Wang SR, Meng W, Jin XC, Zheng BH, Zhang L, Xi HY (2015) Ecological security problems of the major key lakes in China. Environ Earth Sci 74(5):3825–3837
54. Wang T, Zhang M, Springer CH, Yang C (2021b) How to promote industrial park recycling transformation in China: an analytic framework based on critical material flow. Environ Impact Assess Rev 87:106550
55. Wang Y, Zhang H, Zhang C, Liu C (2021a) Is ecological protection and restoration of full-array ecosystems conducive to the carbon balance? a case study of Hubei Province, China. Technol Forecast Soc Chang 166:120578
56. Wang Y, Zhang Z, Chen X (2022a) The Dominant driving force of forest change in the Yangtze River Basin, China: climate variation or anthropogenic activities? Forests, 13(1):82
57. Wei W, Gao Y, Huang J, Gao J (2020) Exploring the effect of basin land degradation on lake and reservoir water quality in China. J Clean Prod 268:122249
58. Wen Z, Hu Y, Lee JCK, Luo E, Li H, Ke S (2018) Approaches and policies for promoting industrial park recycling transformation (IPRT) in China: practices and lessons. J Clean Prod 172:1370–1380
59. Wu HP, Chen J, Zeng GM, Xu JJ, Sang LH, Liu Q, Dai J, Xiong WP, Yuan Z, Wang YQ, Ye SJ (2020) Effects of early dry season on habitat suitability for migratory birds in China's two largest freshwater lake wetlands after the impoundment of three gorges dam. J Environ Inform 36(2)
60. Wu M (2020) An evaluation and spatial statistical analysis of China's regional sustainable development level. Resour Environ Econ 2(1):112–125
61. Xing J, Silva JBD, De Almeida ID (2017) Interest, design and assessment of Eco-Industrial Parks in China within a circular economy paradigm. In: Interest, design and assessment of eco-Industrial parks in China within a circular economy paradigm, pp 12–25
62. Xu J (2018) Research on the resident perception rating for green building through ESGB in China (Master's thesis, University of Twente, Enschede, the Netherlands)

63. Yeung YM, Shen J (eds) (2008) The pan-pearl river delta: an emerging regional economy in a globalizing China. Chinese University Press
64. Yi M, Wang Y, Yan M, Fu L, Zhang Y (2020) Government R&D subsidies, environmental regulations, and their effect on green innovation efficiency of manufacturing industry: evidence from the Yangtze River economic belt of China. Int J Environ Res Public Health 17(4):1330
65. Ying LG (2000) Measuring the spillover effects: some Chinese evidence. Pap Reg Sci 79(1):75–89
66. Yu Q, Li M, Li Q, Wang Y, Chen W (2022) Economic agglomeration and emissions reduction: does high agglomeration in China's urban clusters lead to higher carbon intensity? Urban Clim 43:101174
67. Zhang, X. (2021). Promoting ecological progress with, procuratorial, services. In Beautiful China: 70 Years Since (1949) and 70 People's views on eco-civilization construction. Springer, Singapore, pp 631–643
68. Zhang Y, Zhang J, Wang C, Cao J, Liu Z, Wang L (2017) China and trans-pacific partnership agreement countries: estimation of the virtual water trade of agricultural products. J Clean Prod 140:1493–1503
69. Zheng Y, Xue T, Zhang Q, Geng G, Tong D, Li X, He K (2017) Air quality improvements and health benefits from China's clean air action since 2013. Environ Res Lett 12(11):114020
70. Zheng Z, Qingyun H (2021) Spatio-temporal evaluation of the urban agglomeration expansion in the middle reaches of the Yangtze River and its impact on ecological lands. Sci Total Environ 790:148150
71. Zhong M, Wang L (2020) ecological compensation mechanism of ambient air quality: a case study of Hubei Province, China. Nat Environ Pollut Technol 19(2):663–668
72. Zhou K, Wu J, Fan J, Liu H (2022) Drivers of regional environmental pollution load and zoning control: a case study of the Yangtze River economic Belt, China. Chin Geogr Sci 32(1):31–48
73. Zhou Q, Niu J (2015) A study on the coordinated development of the urban areas around the Qinba mountains. Strat Study Chin Acad Eng 18(5):10–16
74. Zhu M, Shao L (2020) An analysis on the economic cooperation and the industrial synergy of the main river region: from the perspective of the Yangtze river economic zone. J Ambient Intell Humaniz Comput 11(3):1055–1064

Chapter 8
Shandong Byland City Cluster: From an Agricultural Hub to Three Economic Circles and Industrial Clusters

8.1 A General Introduction to the Shandong Byland City Cluster

The Shandong Byland city cluster is located in the eastern part of Shandong Province, including Qingdao, Yantai, Weihai and Weifang, Rizhao, Dongying, and most or part of the prominent Yellow Sea coastal areas, including Bo Hai, Huanghai (east of Penglai) and Liaodong Peninsula [40]. The city cluster includes two sub-provincial cities, Jinan and Qingdao, and six prefecture-level cities, including Zibo, Weihai, Yantai, Dongying, Weifang, and Rizhao. In addition, its jurisdiction covers 22 count-level cities, including Zhangqiu, Jiaozhou, Jimo, Pingdu, Jiaonan, Laixi, Longkou, Leyang, Laizhou, Penglai, Zhaoyuan, Qixia, Haiyang Wendeng, Rongcheng, Rushan, Green State, Zhu City, Shouguang, Anqiu, Gaomi, and Changyi [21]. The total area of the Shandong Peninsula city cluster is 73,000 square kilometers. In 2015, the resident population of the Shandong Byland city cluster was just under 100 million people (at 98,471,600), accounting for 7.2% of China's total. At the same time, the city cluster's GDP was 6.3 trillion Yuan, accounting for 9.3% of the country's total. In 2020, the cluster's GDP rose to 7.3 trillion Yuan [62], but with a lowered proportion of the national total compared to five years before.

According to a report by China [9], the city cluster development plan is to aid the growth of the Shandong province:

> East China's Shandong province recently launched guidelines to facilitate the development of larger city clusters in the province in order to boost regional cooperation and coordinated development. The province aims to build a global influential urban agglomeration on the Shandong Peninsula, foster Qingdao, and Jinan into "national center cities" and cultivate three economic circles.

According to the plan, there are three major economic circles at the provincial level. As it is understood, all three sit under the overarching umbrella of the Shandong Byland city cluster. The first is the capital economic circle, including the provincial capital and six neighbouring cities. This economic circle aims to "*build a demonstration area for ecological protection and a high-quality development in the Yellow*

A. Cheshmehzangi and T. Tang, *China's City Cluster Development in the Race to Carbon Neutrality*, Urban Sustainability, https://doi.org/10.1007/978-981-19-7673-5_8

River Basin, a regional pilot area for the conversion of growth drivers, and an example in cultural cooperation and exchange" [9]. The second economic circle includes five cities of Qingdao, Rizhao, Weifang, Weihai, and Yantai. All these five cities are regarded as prominent "*shipping and trade center, financial center, and a demonstration area for the marine economy*" (ibid). The third economic circle is in the southern areas including four cities of Heze, Jining, Linyi, and Zaozhuang. The plan is to develop this economic cluster or circle as a "*pilot area for rural revitalisation and transformation development and an economic belt in the Huaihe River Basin*" (ibid). While the three proposed economic circles are distinct, they play their parts individually and collectively in the development of the Shandong Byland city cluster. Aside from this economic cluster, the proposal for the Shandong Peninsula Megalopolis, which includes Jinan Metropolitan Area and Qingdao Metropolitan Area, is ongoing. The proposal consists of 11 cities with a total population of 28 million. This regional integration plan aims to promote communication and transportation links, enhancing the geographical distribution of industrial development, institutional development, and economic structure.

The development objective of the Shandong Peninsula City cluster (or Shandong Byland) is to build a transnational urban corridor comprising the Shangdong Peninsula, the Southwest coast of South Korea, and the island of Kyushu in Japan. This development has regional development plans at the international level. It helps to promote the formation of the "Growth Triangle in the Yellow Sea Region between South and North Korea", which goes back to years ago and before the city cluster development plan [1, 18, 19, 29, 30, 44]. At the national level, it is the economic center and leading region of China's Yellow River Basin and one of the growth poles of North China and East China that are on par with the Pearl Delta, Yangtze Delta, and Beijing-Tianjin-Hebei [10, 41]. Together with the Beijing-Tianjin-Hebei, Liaoning, and the region's south [47], the Shandong Byland city cluster aims to jointly build China's Bohai Sea economic cooperation zone leader.

From the national perspective, the construction and development of the Shandong Peninsula city cluster will strategically support cooperation in the Bohai Rim region. Some of this cooperation is related to China's low carbon development [7, 57], port investment and development as part of the Belt and Road Initiative (BRI) [2], and creating multi-level governance to formulate regional brand identities [34]. The city cluster aims to lead the Yellow River Basin to accelerate its development, docking Beijing-Tianjin-Hebei, Yangtze Delta city cluster [6, 8, 43, 52, 68], and optimising the spatial layout of urbanization in China's eastern coastal areas.

From the point of view of the whole province, it is vital to enhance the level of open competition in the Shandong Peninsula city cluster [13]. This proposal is conducive to improving the urban system of the province and promoting the coordinated development of urban and rural areas [5, 16, 37, 38, 63, 65]. It is also conducive to enhancing innovation vitality and promoting industrial structure transformation and upgrading [3, 53, 60]. Moreover, it is conducive to improving resource utilisation efficiency and accelerating the overall planning and green development of land and sea [17, 31]. The city development plan is beneficial to integrate into the national-level regional development strategy and build a new pattern of opening up to the outside world. This approach could lead to a balanced regional development [15, 25]. It is conducive to improving the living standards and quality of urban and rural residents [41] so that they can share the fruits of development and create a new situation of building a solid economic and cultural province.

8.2 A Micro-historical Analysis of the Shandong Byland City Cluster

Before the reform and opening up, the spatial structure of the city cluster in the Shandong Peninsula was slightly different. It includes the Laizhou port, Dengzhou port, and Mizhou port in ancient times [35, 45, 50, 70]. It has been a major maritime trade port between China, the North Korean Peninsula, and Japan [26, 59]. There were two main routes along the ancient Silk Road in China, of which the maritime routes set sail from the eastern part of the Shandong Peninsula [12, 24]. For instance, Langya, along the coast of Jiaozhou in Qingdao, was the earliest maritime port in northern China [71, 72]. Before and after the Sino-Japanese War of 1894–1895 in the Qing Dynasty, many trading ports in the Shandong Peninsula and Korean Peninsula were closed, and the maritime trade and exchange were deserted [36]. Since the 1980s, with the restoration and the improvement of the relations between China and South Korea, and Japan, the port trade in the Shandong Peninsula began to flourish.

The region has also experienced a range of spatial structure formation and reformation. The evolution of the spatial structure of the city cluster in the Shandong Peninsula since reform and opening up, and particularly in the last three decades, suggests the importance of this region at multiple levels. Since the reform and opening up almost 30 years ago, the city cluster of Shandong Peninsula has begun the system of "*changing county from county to city, with city leading county*" to give full play to the central radiation role of the city and meet the requirements of socialist modernisation [41]. Since the 1990s, various transportation hubs in the Shandong Peninsula have been gradually established and perfected [22, 51]. More importantly, industrial and population agglomerations have basically formed the necessary conditions and geographical advantages for the development of city cluster.

8.2.1 An Overview of Population Structure of the Shandong Byland City Cluster

There are 56.14 million urban populations in Shandong Peninsula/Byland city cluster. The urbanisation rate of the permanent population and household registration population have reached 57.01% and 47.5%, respectively, which has entered the middle and late stage of rapid urbanisation [41]. During the "13th Five-Year Plan" period,

17 provinces, autonomous regions, and municipalities have become areas of population outflow, of which 4 have more than 500,000 population outflow [21]. Specifically, Shandong's population outflow is the highest, approaching 900,000 from 2016 to 2019. In 2017, Shandong's resident population exceeded 100 million, reaching 10,058,300. There is an increase of 591,900 from 99,466,400 in 2016. However, in 2017, Shandong's natural population growth reached 1,011,600. In other words, more than 400,000 people left Shandong that year [21]. Data from 2019 shows that the proportion of people over 65 in Shandong reaches 15.84%, and Shandong has entered a deeply aging society. In this regard, we can see the trend of population structure has been affected negatively in recent years.

The 21st Century Economic Research Institute believes that the overall decline of the floating population depends on the reduction of the natural growth rate and the increase of the aging population. Since the 13th Five-Year Plan, especially since 2017, many places have relaxed the conditions for settling in new cities or regions. This also follows the plans under the new-type urbanisation plan [5], which aim to reduce obstacles to labour mobility and population inflows to cities and economic hubs. On the morning of March 4, 2021, at a press conference held in Shandong Province, Shandong announced that it would fully liberalise the urban settlement restrictions to promote two-way population flow. The impacts are yet to be assessed.

8.3 Industrial Structure of the Shandong Byland City Cluster

The Shandong Peninsula has a long history of agriculture in China. It has the highest rate of arable land in China. Thus, it is considered China's major agricultural province, and its agricultural value added has long been ranked first in Chinese provinces [4, 56, 76]. Shandong is not only rich in plant cultivation, livestock, and poultry species resources but also rich in the use of wild animals and plant resources [41]. In addition, Shandong's industrial development, industrial output value, and industrial added value are within the top three provinces in China. Mainly some large enterprises have emerged in the region, known as "group-like economy", and literally translated to industrial clusters [32, 66, 67].

Furthermore, Shandong is an important producer of grain, cotton, oil, meat, egg, and milk in China. Thus, the light industry (and, to some extent, light manufacturing), especially the textile and food industry [28, 49, 54], is quite developed. On the other hand, the development of heavy industry enterprises is rapid. The key industrial and mining enterprises include Qilu Petrochemical Company, Shandong Electric Power Co., Ltd., Shangdong Iron & Steel Co. Ltd., Shengli Oilfield, Yankuang Group, Chalco Ltd., Nanshan Corporation, Chenming Paper, and so on. Lastly, the tertiary industry in the Shandong Peninsula is growing rapidly.

8.4 Overview of the Sustainability Agenda for the Shandong Byland City Cluster

The Shandong Byland city cluster's sustainable development ideas are quite diverse. The governmental stakeholders plan to adhere to the principle of sustainable development, do a good job in urban planning, and promote the coordinated development of population, resources, and the environment [41]. They aim to integrate natural, economic, and social benefits. A key factor is considering urban construction and planning first and as the backbone of ongoing regional development [41]. The primary task of urban planning is to optimise the allocation of urban land resources, improve and strengthen land management, and solve the contradiction of urban construction land [3, 23]. The government aims to gradually improve policies on land replacement, rural homestead management, land use for township enterprises, collective land transfer management, and land use management in development zones [41]. All these sustainability targets are to achieve the goal of rational land use and environmental protection.

In the aspect of population management, measures should be taken to improve population quality, introduce talents, promote a two-way flow of population, and deal with the aging of the population [41]. The overarching aim is to improve education and old-age security systems [11, 20, 69]. The government proposed to strengthen the management of the urban floating population, improve the management of the urban resident population, and form a new model with community organisations as the main body and local government participation [41]. To achieve the sustainability agenda at the regional level, the governments focus on resources and environment, the development of a circular economy, improvement of environmental management [14, 39, 74], and actively guiding the development and utilisation of new materials, new energy, and new technologies.

8.4.1 Sustainable Development Plan from the Environmental Perspective

The government firmly fosters the idea that clear waters and green mountains are mountains of gold and silver [58]. The plan is to make overall planning for the development of regional environmental infrastructure, focus on water, air and soil, improve relevant environmental standards, strictly enforce environmental access rules, and effectively solve environmental pollution problems [41]. The overarching plan is to strengthen regional environmental governance, divided into three sub-strategies as shown below:

(1) **Improving regional water environment quality**
The aim is to enhance water supply capacity, strengthen desalination, and achieve the comprehensive utilisation of seawater. To achieve these, the engaged stakeholders must increase the use of urban reclaimed water and rainwater and

strengthen groundwater overdraft area control. The plan is to jointly build a flood control and disaster mitigation system [41]. The government has already accelerated the construction of a flood control engineering system based on reservoirs, river courses, flood storage, detention areas, and a dam-proof engineering system with sea walls as the main body.

(2) **Coordinated prevention and controlof air pollution**
A few factors have been considered, including strict implementation of the environmental capacity control system and strict examination and approval of air pollution projects. The government aims to establish an exit mechanism for excess capacity in industries with high pollution and high energy consumption, strengthen control of waste gas pollution in thermal power, iron and steel, cement, and petrochemical industries, and intensify efforts to clean bulk coal [41]. Much of the attention is given to strengthening the prevention and control of motor vehicle exhaust pollution and banning the burning of grass and straw. Other secondary plans are to strengthen the catering industry and outdoor barbecue smoke control, impose the strict implementation of dust pollution prevention and control management, and strengthen construction sites, slag transportation, sanitation, cleaning, and other vital links management [41]. The overarching plan is to speed up the urban bare land greening, improve road mechanisation cleaning rate, and control road dust.

(3) **Strengthen soil pollution prevention and control**
The government plans to implement an action plan for the prevention and control of soil pollution, give priority to protecting the soil environment of cultivated land, and crack down hard on the illegal transfer of pollutants from industrial and mining enterprises to the soil [41]. In doing so, it also promotes waste incineration to prevent secondary pollution. The suggested strategies are proposed to improve pollution prevention and control of heavy metals, electronic waste, hazardous chemicals, persistent organic pollutants, and hazardous waste. In recent years, the government also accelerates the construction of hazardous waste disposal facilities and promoted the extension of centralized medical waste disposal to rural medical and health institutions [41]. The secondary strategies are to strengthen the safety management of radioactive sources, advance radioactive waste management, and improve nuclear safety management capabilities.

8.4.2 Sustainable Development Plan from the Economic Perspective

Relying on Shandong Peninsula's coastal geographical advantages is quite important for the region's economic development. It vigorously develops the production and processing of seafood, mainly including the health of seafood culture and seafood processing. The economic clustering of industries helps to achieve the goals through longer-term and larger-scale plans. The government also proposes to prevent over-fishing, strengthen fishing restrictions, scientific development, and utilisation of marine resources, to ensure the sustainable development of marine resources is in place [41]. The government also aims to promote fishery development in the Bohai Sea and Yellow Sea areas [42, 75]. The introduction of domestic and foreign advanced technology and the development of the ship construction and maintenance industry are also key priorities for long-term regional development [41]. The government also pays attention to the utilisation of new energy and new materials, abandons the serious pollution of traditional energy and traditional materials, and achieves the dual objectives of ecological protection and economic development [46]. The Shandong Byland city cluster has a beautiful natural environment and rich coastal resources. It is also a thriving economic center (Yan et al. 2020) to create beach-based scenic spots, and vigorously develop tourism while limiting the number of tourists to prevent excessive development and excessive consumption caused by environmental pollution [41]. The overarching aim is to realise the sustainable development of the ecological environment and tourism.

8.4.3 Sustainable Development Plan from the Social Perspective

The government plans to vigorously develop a circular economy and build a conservation-oriented society [41]. In order to realise the sustainable development of the Shandong Byland city cluster, the plan is to save resources and reduce energy consumption, according to the three principles of circular economy development: reduction, reuse, and recycling. So far, the government has proposed establishing a plan for a circular economy in the short, medium, and long terms. As mentioned earlier, there are also three sub-cluster development strategies as regional economic clusters, which are aimed at promoting population inflows, labour mobility, and social infrastructure development. The establishment of three major industries coordinated the development of the industrial model [41]. The first industry is scientific development, which should nurture education and local industries. The secondary industry mainly focuses on energy utilisation, strengthening new energy and materials guidance, and reducing environmental pollution. Third, the development of the tertiary industry is to focus on consumer areas and the establishment of a sound renewable resources recovery system and resource recycling mode.

8.5 Pathways to Achieve China's Carbon Neutral Plan: Opportunities and Challenges

The Shandong Byland city cluster has a long way to go to achieve carbon neutrality targets. Compared to earlier city clusters that we have mapped in this book, the Shandong Byland city cluster is still at the earlier stage of industrial restructuring. One prominent plan is to evolve the regional power grid base to achieve carbon neutrality targets [64]. The allocation of nuclear plants and renewable energy capacity for power generation is the fundamental strategy to satisfy carbon neutrality requirements. Another promising approach is to attain the industrial water-use target [33], benefitting from long coastal lines in the region. However, the current spatial structure and industrial distribution are the bottlenecks to such targets.

In addition, as [61] highlighted, Shandong Province has one of the highest values of marine carbon sink. Yet, the capacity is not fully utilised to achieve the carbon neutrality targets on carbon emission reductions and management. For now, a total-factor CO_2 emissions efficiency measure is proposed to achieve the emission reduction targets [27]. This is specifically relevant to industrially developed regions, where regional development is still ongoing. The Shandong Byland city cluster is under significant pressure to reduce its carbon allowances [48], indicating the pressure that is mainly on heavy industries and industrial clusters. Thus, we expect to see shifts in energy transfer networks at the regional level, agricultural production, and heavy industries that could help achieve the carbon neutrality goal of 2060. Prominent proposals like centralised Biogas [55], nuclear power generation, and marine-based strategies could help improve the regional and provincial disparities in energy systems, CO_2 emissions, and cost-effectiveness [73]. Thus, we see a relatively more difficult path ahead for the Shandong Byland city cluster, yet with some promising strategies that could make the economic circles into global models of carbon neutral industrial hubs.

References

1. Cossa RA, Khanna J (1997) East Asia: economic interdependence and regional security. Int Aff 73(2):219–234
2. Chen J, Fei Y, Lee PTW, Tao X (2019) Overseas port investment policy for China's central and local governments in the belt and road initiative. J Contemp China 28(116):196–215
3. Chen K, Long H, Liao L, Tu S, Li T (2020) Land use transitions and urban-rural integrated development: theoretical framework and China's evidence. Land Use Policy 92:104465
4. Chen W, Wu S, Lei Y, Li S (2017) China's water footprint by province, and inter-provincial transfer of virtual water. Ecol Ind 74:321–333
5. Cheshmehzangi A (2016) China's New-Type Urbanisation Plan (NUP) and the foreseeing challenges for decarbonization of cities: a review. Energy Procedia 104:146–152
6. Cheshmehzangi A, Dawodu A, Sharifi A (2021) Sustainable urbanism in China. Routledge, New York

7. Cheshmehzangi A, Li H, Yang X (2018) Low carbon and climate resilient urban development in China. ADB Report, developed for the National Development Reform Committee (NDRC), Beijing, China
8. Cheshmehzangi A, Li Y, Li H, Zhang S, Huang X, Chen X, Su Z, Sedrez M, Dawodu A (2022) A hierarchical study for urban statistical indicators on the prevalence of COVID-19 in Chinese city clusters based on multiple linear regression (MLR) and polynomial best subset regression (PBSR) analysis. Sci Rep 12(1):1–16
9. China Daily (2020) City clusters to aid growth in Shandong. Available from: http://shandong.chinadaily.com.cn/2020-06/09/c_499045.htm
10. CRE (2017) Shandong issued the "Shandong Peninsula urban agglomeration development plan" to clarify the "Four Positions". <山东印发《山东半岛城市群发展规划》明确"四大定位">. Available from: http://www.cre.org.cn/list2/cz/11487.html
11. Cui J (2004) Countermeasures for implementing sustainable development strategy in Shandong, China. Chin J Popul Resour Environ 2(4):57–60
12. Dong X (2019) The Maritime Silk Road. In: Fang C, Nolan P (eds) Routledge handbook of the belt and road. Routledge, New York, pp 68–73. Available from: https://www.taylorfrancis.com/chapters/edit/https://doi.org/10.4324/9780429203039-13/maritime-silk-road-dong-xiangrong
13. Douding Architecture (2010) Analysis on the evolution of spatial structure of Shandong Peninsula urban agglomeration _ 926 <浅析山东半岛城市群空间结构的演变_926>. Available from: https://jz.docin.com/p-52421244.html
14. Fan Y, Fang C (2020) Circular economy development in China-current situation, evaluation and policy implications. Environ Impact Assess Rev 84:106441
15. Fang C (2021) The Bo-Tai Line: establishing the concepts of a balanced regional development line and a national development backbone. J Geog Sci 31(6):839–858
16. Fang X, Shi X, Phillips TK, Du P, Gao W (2021a) The coupling coordinated development of urban environment towards sustainable urbanization: an empirical study of Shandong Peninsula, China. Ecol Indic 129:107864
17. Fang Y, Zhai T, Zhao X, Chen K, Guo B, Wang J (2021) Study on the comprehensive improvement of ecosystem services in a China's bay city for spatial optimization. Water 13(15):2072
18. Hong S, Khim JS, Naile JE, Park J, Kwon BO, Wang T, Lu Y, Shim WJ, Jones PD, Giesy JP (2012) AhR-mediated potency of sediments and soils in estuarine and coastal areas of the Yellow Sea region: a comparison between Korea and China. Environ Pollut 171:216–225
19. Hook GD (2002) The Japanese role in emerging microregionalism: the Pan-Yellow Sea economic zone. In: Microregionalism and world order . Palgrave Macmillan, London, pp 95–114
20. Hu Y, Ren M (2019) Research on the intelligent old-age care mode based on the combination of medical and health care under the perspective of" Internet+". In: 2019 5th international conference on social science and higher education (ICSSHE 2019). Atlantis Press, pp 439–442
21. IFENG (2021) 21st Century Business Herald "Thirteenth Five-Year Plan" population migration map: population outflow of 17 provinces, nearly 5 million people to Guangdong, Zhejiang 十三五"人口迁徙图谱: 人口流出省份17个, 近500万人流向广东、浙江. Available from: https://finance.ifeng.com/c/84YVbB2xXGo
22. Jin F, Ding J, Wang JE, Liu D, Wang C (2012) Transportation development transition in China. Chin Geogra Sci 22(3):319–333
23. Jin G, Chen K, Wang P, Guo B, Dong Y, Yang J (2019) Trade-offs in land-use competition and sustainable land development in the North China plain. Technol Forecast Soc Chang 141:36–46
24. Kang H (2013) The spread of Sarnath-Style Buddha images in Southeast Asia and Shandong, China, by the sea route. KEMANUSIAAN: Asian J HumIties 20(2)
25. Kang J, Zhang X, Zhu X, Zhang B (2021) Ecological security pattern: a new idea for balancing regional development and ecological protection. a case study of the Jiaodong Peninsula, China. Glob Ecol Conserv 26:e01472
26. Kim HM (2021) Inter-continental transport networks and Asian Economic Corridor for the Korean Peninsula. International perspectives on the belt and road initiative. Routledge, New York, pp 100–118

27. Li S, Diao H, Wang L, Li L (2022) A complete total-factor CO_2 emissions efficiency measure and "2030–60 CO_2 emissions targets" for Shandong Province, China. J Clean Prod 132230
28. Li W, Fan J, Qu T, Zhao L, Tong Q, He Y (1992) The industrial structure and distribution of the Bohai-Yellow Sea Rimland of China in the 1990s. Chin Geogra Sci 2(2):97–113
29. Lim SH (2015) How beneficial would the construction of a Rason-Hunchun sub-regional economic cooperation zone in the Northeast Asian Borderlands Be? North Korean Rev 63–81
30. Lim SH, Lim KT (2006) Special economic zones as survival strategy of North Korea. North Korean Rev 47–61
31. Liu J (2012) Study on sustainable development of ecological environment in Shandong Peninsula urban agglomeration 山东半岛城市群地区生态环境可持续发展研究. From Shandong University, Available from: https://www.docin.com/p-2095051390.html
32. Liu L (2008) Local government and big business in the People's Republic of China-case study evidence from Shandong Province. Asia Pac Bus Rev 14(4):473–489
33. Liu X, Dai H, Wada Y, Kahil T, Ni J, Chen B, Chen Y, Guo C, Pan C, Liu X, Liu Y (2022) Achieving carbon neutrality enables China to attain its industrial water-use target. One Earth 5(2):188–200
34. Lu H, de Jong M, Song Y, Zhao M (2020) The multi-level governance of formulating regional brand identities: evidence from three Mega City Regions in China. Cities 100:102668
35. Ma, G. (2016). *The Shandong Peninsula in East Asian Maritime History during the Yuan-Ming Transition* (Doctoral dissertation, Ghent University).
36. Olender P (2014) Sino-Japanese Naval war 1894–1895, vol 3105. MMP Books, Kerala
37. Qiao X, Meng Y, Zhu X (2015) Research on the hierarchical structure of the metropolitan system and economic development of the provincial capital urban circle in Shandong. Chin J Popul Resour Environ 13(1):78–86
38. Qu Y, Zhan L, Jiang G, Ma W, Dong X (2021) How to address Population Decline and Land Expansion (PDLE) of rural residential areas in the process of Urbanization: a comparative regional analysis of human-land interaction in Shandong Province. Habitat Int 117:102441
39. Ren Q (2011) Circular economy action programs and countermeasures for small and medium-sized resource-based cities of China-case study of Zibo City of Shandong Province. Energy Procedia 5:2183–2188
40. Shandong Government (2012) Drafted report on: sustainable Development path of urbanization in Shandong Peninsula urban agglomeration. 山东半岛城市群城市化的可持续发展道路. Available from: https://www.docin.com/p-503596046.html
41. Shandong Government (2021) Shandong Peninsula urban agglomeration development plan (2021–2030) 山东半岛城市群发展规划(2021–2030年). Available from: http://www.shandong.gov.cn/art/2022/1/6/art_107861_116576.html
42. Song J, Duan L (2019) The Bohai Sea. In: World Seas: an environmental evaluation. Academic Press, Washington D.C., pp 377–394
43. Song X, Wu S, Xu X (2019) New trends of regional development since the 18th CPC national congress. In: The great change in the regional economy of China under the new normal. Palgrave Macmillan, Singapore, pp 161–191
44. Sum NL (2008) Book review: as borders bend: transnational spaces on the Pacific Rim. In: Chen X, Lanham MD (esd) Rowman and littlefield, vol 331, pp. 89.00(cloth); 34.95. Urban Aff Rev 43(3):430–432
45. Sun GQ, Hugh I (1996) A reserch into the sea route between Silla and Tang from the 8th Cent. AD to the CENT AD. In: Proceedings of the Korean institute of navigation and port research conference . Korean Institute of Navigation and Port Research, pp 21–29
46. Sun Y (2020) The achievement, significance and future prospect of China's renewable energy initiative. Int J Energy Res 44(15):12209–12244
47. Tao Y (2022) Comparison of three bay area economic belts and the position and role of Shenzhen in the Guangdong-Hong Kong-Macao Greater Bay Area. In: Annual report on the development of China's special economic zones. Springer, Singapore, pp 1–9
48. Tian M, Hu YJ, Wang H, Li C (2022) Regional allowance allocation in China based on equity and efficiency towards achieving the carbon neutrality target: a composite indicator approach. J Clean Prod 342:130914

49. Tian P, Lin B (2018) Regional technology gap in energy utilization in China's light industry sector: non-parametric meta-frontier and sequential DEA methods. J Clean Prod 178:880–889
50. Vivier BT (2008) Chinese foreign trade, 960–1276. Yale University. Available from: https://www.proquest.com/intermediateredirectforezproxy
51. Wang C, Wang J (2011) Spatial pattern of the global shipping network and its hub-and-spoke system. Res Transp Econ 32(1):54–63
52. Wang JA, Liang S, Shi P (2022a) Land Cover and Land Use. In: The geography of contemporary China. Springer, Cham, pp 121–152
53. Wang J, Wang C, Yu S, Li M, Cheng Y (2022) Coupling coordination and spatiotemporal evolution between carbon emissions, industrial structure, and regional innovation of counties in Shandong Province. Sustainability 14(12):7484
54. Wang L (2011) Development of circular economy and optimization of industrial structure for Shandong Province. Energy Procedia 5:1603–1610
55. Wang Q, Yu L, Yang Y, Zhao H, Song Y, Song W, Liu J (2022) Let the farmers embrace "Carbon Neutrality": taking the centralized biogas as an example. Int J Environ Res Public Health 19(15):9677
56. Wang W, Gao L, Liu P, Hailu A (2014) Relationships between regional economic sectors and water use in a water-scarce area in China: a quantitative analysis. J Hydrol 515:180–190
57. Wang Y, Lan Q, Jiang F, Chen C (2019) Construction of China's low-carbon competitiveness evaluation system: a study based on provincial cross-section data. Int J Clim Chang Strat Manag 12(1). Available from: https://doi.org/10.1108/IJCCSM-11-2018-0076/full/html
58. Wang Z, He C, Zhang S (2020) Eco-civilization and Sustainable Development: the case of China's countryside. Sustain Environ Decis Mak 1–21
59. Wei C (2005) The constitution of cultural routes between China and the North East Asia during Han/Tang Dynasties and their protections. ICOMOS Open Archive. Available from: http://openarchive.icomos.org/id/eprint/460/
60. Wu N, Liu Z (2021) Higher education development, technological innovation and industrial structure upgrade. Technol Forecast Soc Chang 162:120400
61. Wu X, Tian Z, Guo J (2022) A review of the theoretical research and practical progress of carbon neutrality. Sustain Oper Comput 3:54–66
62. Xinhua Silk Road (2022) Understanding Shandong: plan for city cluster on Shandong Peninsula unveiled to boost regional development. Available from: https://en.imsilkroad.com/p/325808.html
63. Yanbo Q, Guanghui J, Yaya T, Ran S, Shuwen W, Yuling L (2019) Urban-Rural Construction Land Transition (URCLT) in Shandong Province of China: features measurement and mechanism exploration. Habitat Int 86:101–115
64. Yang L, Tian X, Zhang D, He H, Li K, Guan X (2022) Study on evolution path of Shandong power grid based on" Carbon Neutrality" goal. In: *2022* 5th international conference on energy, electrical and power engineering (CEEPE). IEEE, pp 770–775
65. Yang Z, Shen N, Qu Y, Zhang B (2021) Association between rural land use transition and urban–rural integration development: from 2009 to 2018 based on county-level data in Shandong Province, China. Land 10(11):1228
66. Yu C, Liu H (2013) Current situations and measures of regional brand spreading in China—take Shandong as an example. In: 2nd international conference on management science and industrial engineering (MSIE 2013). Atlantis Press, pp 574–577
67. Yuan G (2016) Small farming in the market economy: a study of a village in Shandong, and its theoretical significance. In: Mapping China. Brill, pp 7–35
68. Yuan K, Hu B, Niu T, Zhu B, Zhang L, Guan Y (2022) Competitiveness evaluation and obstacle factor analysis of urban green and low-carbon development in Beijing-Tianjin-Hebei Cities. Math Probl Eng. Available from: https://www.hindawi.com/journals/mpe/2022/5230314/
69. Yuan S, Wang Y-Y (2018) Difficulties and solution of centralized support for rural five guarantees—based on the investigation of the elderly nursing home in shandong province. J Hebei Univ (Philosophy and Social Science) 43(4):126

70. Yuan X (1997) On the Chinese and Korea ancient ships in the 9th century and the historical remaining of xinluos living in China. In: Proceedings of the Korean institute of navigation and port research conference. Korean Institute of Navigation and Port Research, pp 223–233
71. Zhang G (2015) Research on safety management for marine tourist ships in Jiaozhou Bay. Master dissertation at Dalian Maritime University. Available from: https://commons.wmu.se/cgi/viewcontent.cgi?article=1131&context=msem_dissertations
72. Zhang M, Rasiah R (2013) Qingdao. Cities 31:591–600
73. Zhang R, Hanaoka T (2021) Deployment of electric vehicles in China to meet the carbon neutral target by 2060: provincial disparities in energy systems, CO_2 emissions, and cost effectiveness. Resour Conserv Recycl 170:105622
74. Zhang X, Zhao-Hui Z, Zhang X, Yao H, Zheng K (2013) Developmental mode for circular economy in the Yellow River Delta. Int J Environ Sci Dev 4(6):662
75. Zhou Q, Zhang H, Fu C, Zhou Y, Dai Z, Li Y, Tu C, Luo Y (2018) The distribution and morphology of microplastics in coastal soils adjacent to the Bohai Sea and the Yellow Sea. Geoderma 322:201–208
76. Zou L, Liu Y, Wang Y, Hu X (2020) Assessment and analysis of agricultural non-point source pollution loads in China: 1978–2017. J Environ Manage 263:110400

Chapter 9
Guangdong-Fujian-Zhejiang Coastal Region: A Network Corridor Between Three Coastal Provinces

9.1 A Brief Introduction to the Guangdong-Fujian-Zhejiang Coastal Region

The Guangdong-Fujian-Zhejiang (GFZ) coastal region is a large city cluster stretched between three major eastern and south-eastern coastal provinces in China. Major cities include Wenzhou in the southern part of the Zhejiang Province; in Fujian Province, cities of Ningde, Fuzhou, Putian, Quanzhou, Xiamen, and Zhangzhou; and in Guangdong Province, cities of Shantou, Jieyang, and Chaozhou. Fuzhou and Xiamen are the city cluster's two core cities, both located in the Fujian Province, which sits between the other two provinces. Its strategic location at the side of the Taiwan Strait makes it important for political, cultural, trade, and business exchanges [38]. The GFZ city cluster covers a total area of 270,000 km2 with an estimated population of 56.12 million and accumulated GDP of 4859.88 billion yuan in 2020.

In recent years, with the rapid development of the Yangtze River Delta and Pearl Delta metropolitan areas, the lack of core cities in Fujian is facing the considerable pressure of being siphoned by the two economic regions at both ends [33]. Take Xiamen University, ranked first in Fujian Province, for example. It is recorded that 90% of its graduates leave for the Yangtze River Delta and the Pearl Delta, while less than 5% are willing to work in Xiamen [51]. Another amazing phenomenon originated in Fujian, indicating that Jinyuan Group, Rongxin Group, and Zhengrong Group have moved their headquarters to Beijing, Shanghai, and other places. There was once the headquarters of the "Fujian Company" or "City Company", but they are no longer based in the province. The reason for brain drain and enterprise drain is the lack of metropolis in the region, city group, and economic circle [51]. The primary purpose of planning the coastal urban agglomeration in Guangdong, Fujian, and Zhejiang is to concentrate on the development and construction of major cities such as Wenzhou, Fuzhou, Quanzhou, Xiamen, and Shantou [8, 11, 13, 27, 30]. The formation of a linear urban corridor pattern improves the economic development level of southeast coastal areas [13]. Thus, it is planned to become the transition

A. Cheshmehzangi and T. Tang, *China's City Cluster Development in the Race to Carbon Neutrality*, Urban Sustainability, https://doi.org/10.1007/978-981-19-7673-5_9

zone connecting the Pearl River Delta and the Yangtze Delta to avoid the danger of being drained off by the two places.

The construction of urban agglomerations on the west bank of the Taiwan Strait is an urgent need to develop Fujian Province's comparative advantages further and achieve sound and rapid development [8]. It is a significant move to improve the economic layout of coastal areas and promote the development of the other regions on the west side of the Taiwan straits and areas where Taiwan investment is relatively concentrated [23, 32, 52, 60, 66]. It is also a strategic plan for strengthening cross-strait's exchanges and cooperation and promoting the great cause of the peaceful reunification of China. Thus, for China Mainland, the GFZ city cluster is an essential region for further development of industries, cities, and economic zones. Also, it is commonly known to be of great economic and political significance.

Proceeding from the overall interests of safeguarding the core interests of China, the central government firmly grasps the theme of peaceful development of cross-strait's relations, strive to promote cross-strait exchanges and cooperation [15], and promotes mutual benefit and win–win results [8]. In doing so, regional development focuses on transforming the mode of economic development and strengthening the capacity for independent innovation to improve the quality and level of economic growth [19, 58, 61]. The primary focus is on multiple modes of development, i.e., urban, rural, and regional development. The strategies are set to improve the coordination of economic and social development, which are crucial to China's new-type urbanisation plan (NUP) and urban–rural development directions [3, 4]. Thus, the government focuses on deepening reform and opening up to enhance development's driving force and vitality [50]. In return, the government plans to improve people's livelihood and promote the construction of a harmonious socialist society. These are strategised to strengthen ecological progress, enhance the regional capacity for sustainable development, and build the urban agglomeration on the west bank of the Taiwan Strait into a harmonious region with sustained economic growth, a more prosperous culture, growing comprehensive competitiveness, and happy life for the people [50]. The results are expected to lead to make more outstanding contributions to the overall situation, which has been under the shadow of economic, political, and trade tensions for decades.

9.2 A Historical Analysis of the Guangdong-Fujian-Zhejiang Coastal Region

At the beginning of the nineteenth century, the Chaoshan area began to trade with the outside world, and the status of Shantou port was comparable with that of Guangzhou port [9, 25, 31, 43]. Xiamen is a coastal city in China and one of the first four special economic zones to implement the policy of opening up [22, 34, 44]. It is also a modern international port scenic tourist city, with the first batch of national 5A-class tourist attractions—Gulangyu [6, 54, 59, 65]. Former US President Richard

Nixon once praised Xiamen as the "Oriental Hawaii" [59]. In 1842, after the first Opium War, China signed the "Treaty of Nanjing" with the United Kingdom of Great Britain and Ireland [39], which led to the opening of Shanghai, Guangzhou, Fuzhou, Xiamen, and Ningbo as the treaty ports, known as the five-port trade [1, 16, 21]. It was the beginning of foreign trade in China. In the past 100 years, some of the five port trading cities have developed into world-class cities, such as Shanghai and Guangzhou [49]. Fuzhou, Ningbo, and Xiamen are also important cities in the southeast coastal regions of China. All five cities also made a list in the 2017 ranking of the world's top 100 container ports. Three of the world's top ten seaports and five of the top 100 ports were among the five ports in the world. Specifically, the 2017 ranking indicates Shanghai port (first), Ningbo-Zhoushan port (fourth), Guangzhou port (seventh), Xiamen (14th), and Fuzhou (57th). And if ranked according to the cargo throughput, Ningbo-Zhoushan Port has firmly ranked as the largest port in China and even the world [14, 50]. It is also the world's first annual cargo throughput of over 900 million tons of port [60]. More importantly, the five ports also laid the foundation for developing cities on the west side of the strait.

In 2009, the "Coordinated Development Plan of Urban Agglomerations on the West Coast of the Taiwan Straitwas approved by the state, and the urban agglomeration on the west side of the strait appeared for the first time and rose to a national urban agglomeration [27, 30, 67]. More recently, the urban agglomerations on the strait's west bank span four Fujian, Zhejiang, Guangdong, and Jiangxi provinces, including nine prefectures and cities in Fujian and three in Zhejiang, and four in Jiangxi and Guangdong. In 2021, the People's Republic of China's 14th Five-Year Plan for National Economic and Social Development and the Outline of Long-term Goals to 2035 was officially announced. It was the birth of China's new growth story coupled with the 2060 carbon neutrality pledge [41, 42]. It is worth noting that, in the outline of the 14th Five-Year Plan, the original expression of the city group on the west bank of the strait has disappeared instead of the formulation of the coastal city group of Guangdong, Fujian, and Zhejiang. Yet, after almost two years from the carbon neutrality pledge, the focus of the regional development in this city cluster is still on the coastal areas.

9.2.1 The Significance of the Coastal City Cluster: A Strategic Coastal Network

The significance of upgrading "Urban Agglomeration on the West Bank of the Strait" to "Guangdong, Fujian, and Zhejiang" is critical to current and future economic growth and development plans. As the name implies, the city cluster on the west side of the Strait is the corresponding city group on the east side. Although the urban agglomeration' is one of the most extensive urban agglomerations spanning the provinces, it requires more attention for future development plans. Historically, it has been overshadowed by neighbouring economic hubs in the north and south.

However, for a long time, the development of the city cluster was less than expected, and the degree of integration between several provinces was insufficient. Although Fujian has been in a leading position, it is difficult to really play a leading role in this regional development. Therefore, this time, from the "West Bank City Group" to the "Guangdong, Fujian, and Zhejiang city group" [38] is not just a name change but a considerable span.

The forecasts of the GFZ city cluster development indicate the domination of the Fujian province and the extension of trades with the other side of the StraitStrait. If so, the GFZ city cluster could become a breakthrough for integrating the three provinces' development, focusing on the regional interior. In this pathway, the future is no longer only a growth pole, but a number of central cities and metropolitan areas constitute the first-class urban agglomeration and a significant economic and trade region. Cities in this cluster need further attention for future development. For instance, Shantou is located in Shantou Jie Chao Metropolis Circle, a sub-central city of Guangdong Province. It is also one of the two pillars of the coastal economic belt of Canton (the other is Zhanjiang). Another example is Wenzhou, a city with the third highest GDP in the Zhejiang Province [28, 29, 37]. It is also one of the four metropolitan areas (other than Hangzhou, Ningbo, and Jinyi) with a highly developed private economy. Thus, there is a phenomenon called the 'Wenzhou Model of Development' [12, 18, 37, 40, 48, 56], which resembles China's early stages of modernisation, transnational development [26], industrial development, and rural industrialisation.

Furthermore, Fujian's Fuzhou, Quanzhou, and Xiamen are equally powerful [38]. In 2020, the GDP of Fuzhou and Quanzhou both exceeded one trillion Yuan, and Xiamen exceeded 600 billion Yuan [7]. These three cities are Fujian's high-growth pole. The birth of Guangdong, Fujian, and Zhejiang urban agglomeration means that Fuzhou, Xiamen, Shantou, and Wenzhou will be equally important in the forthcoming regional development plans. Hence, the aim (now in line with the carbon neutrality plan), is to accelerate the development of these regions.

9.3 Industrial Structure in the Guangdong-Fujian-Zhejiang Coastal Region

Due to the historical basis, resource structure, social culture, market basis, local system, and other factors, the cities in the region have formed "overlapping" industries [8]. This industrial formation is especially visible in the manufacturing structure, which indicates an apparent homogeneous competition phenomenon [38]. Because of the region's high convergence of industrial structure, the competitiveness of industries in the area is greater than the complementarity. Similar examples are seen elsewhere, nationally in China [55] and internationally in other cluster or region-industry developments [10]. Such development mode deviates from the laws and objective requirements of the regional division of labour. It damages the reasonable industrial division among regions and sacrifices the comparative benefits between regions, which becomes a significant obstacle to industrial cooperation [7]. Due to the lack of big cities and leading industries, Fujian has not formed a real economic zone [7]. The industrial function is weak, the industrial chain between the cities is short, and the industry groups are almost independent. Thus, the correlation effect is unsatisfactory today. More importantly, it is difficult to form the resource integration effect and interactive development effect [7], which affects the competitiveness of the city groups.

9.3.1 *Summary of Challenges*

The first is the lack of major urban hubs in the region. There is a lack of one or two mega-cities with a population of more than 10 million (or several large cities with a population of 5–10 million), which becomes the lack of a leading radiation role [7]. All three major cities in Fujian, including Fuzhou, Xiamen, and Quanzhou, have less than 5 million urban populations. Therefore, the city's primacy is far from enough [50]. The same goes for Chaoshan, Wenzhou, and Taizhou. The second is the lack of industrial clusters with strong competitiveness at home and abroad. In 2020, Fujian Province had 39 billion Yuan earnings from its industrial clusters. But the variety and number of such industrial clusters are limited. Thus, this limitation could be argued as the second challenge with the urban agglomerations of such a level.

The third challenge is that the GFZ city cluster is not close enough to Taiwan's economy, industry, capital, technology, and talents. Since the reform and opening up, the capital and industrial talents from Taiwan have gone south to the Pearl River Delta of Dongguan and Shenzhen and to the north to Shanghai, Suzhou, and Kunshan [2, 17, 20, 57, 64]. Thus, the GFZ has not yet attracted talent and economic investment compared to its neighboring coastal regions. Furthermore, we can argue the scarcity of factors such as talent, capital, technology, and industry facing the north of the Yangtze River Delta and the south of the PRD siphon effect is critical to the current and future regional development plans. The GFZ city claustra have faced a relatively long time and a large number of brain drain. For example, many headquarters economy and enterprises have already moved to Shanghai or Shenzhen, such as the famous four major real estate companies in Fujian.

9.3.2 *Summary of Coping Strategies*

Here, we summarise coping strategies that are highlighted by the local and provincial governments, mapping what are the factors that could strengthen the GFZ's regional development: The first is to strive to develop one or two mega-cities in the region, including the central government to support the merger of Xiamen and Zhangzhou into the New Xiamen City. So far, New Xiamen City has been covered by policies such as the Pilot Area for Comprehensive Reform of Taiwan in Special Economic

Zones and Free Trade Zones [35, 63]. With this plan, Xiamen could soon become a mega-city with a population of over 10 million and a GDP of over one trillion Yuan in the coastal urban circle of Guangdong Fujian and Zhejiang, striving to become the central city of the country [50]. The second is to improve the software and hardware business environment of Guangdong, Fujian, and Zhejiang coastal urban agglomeration. In doing so, the cooperation should lead to the establishment of an international, legalised, convenient, and market-oriented business environment [50].

The third is the establishment of a number of high-level science and engineering universities, such as the Chinese Academy of Sciences (CAS), Xiamen University of Technology, and so on. Such educational investment is aimed to support regional population growth and personnel training that could be retained for their careers in the future [51, 58]. The ageing population trend and brain drain effects currently put pressure on how education and industry could be linked together in a longer-term plan. Fourth is to make every effort to attract investment. In this approach, the region should aim to take advantage of the convenience of the rich in Southeast Asia and Taiwan, Give play to Taiwan's five edge advantages to strive for China's electronics, petrochemical, and other industries in Taiwan and Guangdong, as well as Fujian and Zhejiang coastal international best port resources [7]. Through this approach, the establishment of a modern high value-added port industry group should be prioritised. The fifth and last is the focus on institutional innovation and institutional changes, such as the farmland property rights system, and household registration system changes, to solve the Guangdong-Fujian-Zhejiang coastal urban agglomeration people, land, and finance issues. The GFZ coastal urban agglomeration, with its core, is the Fuzhou-Xiamen urban agglomeration development quality, determines the quality of the development of the urban agglomeration.

9.4 The Cluster's Sustainability Agenda: Plans and Perspectives

In this section, we summarise the regional sustainable development plans from two primary perspectives environmental and economic. Once again, governmental reports and documents are studied to ensure local and regional plans are collected through adequate knowledge and based on valid sources.

9.4.1 Sustainable Development Plan from the Environmental Perspective

The first aspect is to protect mountains, rivers, and coastal zones and build green corridors along river basins and transport corridors [13]. The government plans to organise a series of ecological resources and green open spaces, and build a "double ring, belt, two rivers, network corridor" as the main regional ecological pattern. In doing so, the plan is to maximise the protection of beautiful mountains and rivers' ecological environment in the region. "Double ring" refers to the circular ecosystem composed of mountains and hills [38]. "The first ring" is for the Wuyi Mountains and Hawksbill Mountains, Nanping, Sanming, and Longyan part of the ecological barrier. The "Second Ring" consists of Taimu Mountain, Jiufeng Mountains, Daiyun Mountain, and Boping Ridge, which is the ecological barrier for cities in the province [7]. As the plans indicate, the forest ecosystem should be protected, and exploitation should be restricted.

On the other hand, "Belt" refers to a coastal zone consisting of the Bay Area, offshore islands, and offshore waters [7]. It is necessary to optimise the use of the inshore/coastal sea area, rationally allocate industrial areas and towns, and consider the relationship between the development and protection of the shoreline as a whole. The water environment quality should be strictly controlled, the relationship between the upstream and the downstream should be coordinated, the ecological compensation method should be improved, and the development of aquaculture and mining (sand) should be controlled strictly [51]. Lastly, the "Network corridor refers to the ecological isolation corridor formed by the rivers of Jinjiang, Tingjiang, AoJiang, Mulan River, Jiaoxi, and the green belts along the river as regional transportation corridors and green belts. The "Two Rivers" refer to the Min River and the Jiulong River, critical ecological corridors in Fujian Province [24, 62].

The second aspect is to consider ecological function area division and zoning management. According to the regional ecological pattern, It is planned to be five ecological functional areas, such as (1) mountain basin and valley ecological area in north and west of Fujian, (2) original ecological area of middle and low mountains in east Fujian and middle of China, (3) coastal area and nearshore sea area in eastern Min, (4) hilly basin in western southeast Min and coastal, and (5) coastal sea areas in southeastern Min [7]. North and West Fujian mountain basin valley ecological areas are of significant importance in the sustainable development plans of the region. This is aligned with protecting water conservation and biodiversity in the headwaters of the Minjiang and Tingjiang rivers. Hence the plans are to protect the ecological environment of important reservoirs in the upper and middle reaches of rivers and to protect the Wuyishan World Cultural and Natural Heritage Site, Taining "China Danxia" natural heritage. The government also pledged to focus on the prevention and control of industrial pollution sources in Yongan-Sanming-Shaxian valley in the middle reaches of Shaxi [7]. Hence, comprehensive control of soil and water loss in the middle and upper reaches of the Tingjiang River is planned for the coming years. We summarise some of these plans here:

The original ecological region of middle and low mountains in eastern Fujian.

To protect the water conservation and biodiversity of the headwaters of the rivers in eastern Fujian, Mulan River, Jinjiang River, and Jiulong River; focus on monitoring and controlling soil erosion; to protect the ecological environment of the Baishui Ocean and Gutian-Shuikou reservoir area, and to strengthen the urban ecological construction in the region.

The coastal and nearshore ecological zones in eastern Fujian.

To protect the marine ecosystem, maintain marine biodiversity, implement marine disaster prevention and control; focus on monitoring ecological security; coordinate the relationship between the construction of harbors, industries, and cities; and the ecological protection of the bay.

Ecological region of hilly basin in western southeast Fujian.

To protect biodiversity, strictly control new orchard and quarry projects, prevent and control soil erosion; accelerate the process of urbanisation, balance urban and rural development; and rational use of land resources, strict control of breeding industry and other non-point source pollution, sheet metal processing, mining pollution, and excessive development of mountain soil erosion.

Southeast Fujian coastal and nearshore ecological zone.

To promote the ecological construction of cities and towns, and build them into national ecological garden model cities and environmental protection model cities; to comprehensively control pollution, strictly control the discharge of pollutants into the bay; to protect the marine ecosystem and maintain the diversity of marine life and estuarine wetlands. Moreover, the scientific demonstration should be carried out for reclamation in the inner bay area to maintain the stability of the harbor and waterway; strengthen the ecological construction of seaside tourist areas, and to maintain and improve the regional ecological carrying capacity.

The third aspect is to consider the importance of the ecological function of the protected areas. In this regard, areas with essential ecological functions in Fujian Province are diverse, such as the headwaters of major rivers, important water conservation areas, estuarine wetlands, key prevention and supervision areas for soil and water conservation, drinking water source protection areas, important fishing areas, and areas with a concentrated distribution of wild animals and plants. Thus, the urgent need to establish ecological function reserves has become visible in recent agendas and plans. The government proposes that protected area preservation shall be implemented in the source areas of rivers, important water source conservation areas, key prevention and supervision areas for water and soil conservation, important fishery waters, and areas with a concentrated distribution of wild animals and plants [7]. The ongoing and future projects are focused on restricting development and construction in fragile ecological environment areas and important ecological function protected areas. Thus, the development and construction are prohibited in nature reserves and areas of special conservation value. The government also proposes to stop all exploitation activities and other artificial destruction activities in the protected Areas that lead to the continued degradation of ecological functions. The total quantity (capacity) control of the breeding industry shall be implemented, and the construction of all projects that produce serious environmental pollution is planned to be stopped [7]. The government also plans to change the extensive production and management mode to ensure the destroyed ecosystem is gradually reconstructed and restored.

Lastly, the government aims to strengthen comprehensive improvement and ecological reconstruction in some areas where the ecological environment is fragile and the trend of degradation is emerging [7]. The overarching plan is to preserve and protect important ecological functions that are implemented at the regional level. The government's forthcoming plan is to build demonstration zones for ecological restoration in key mining areas such as the Zijinshan Mine in Shanghang and establish demonstration zones in eastern Fujian for comprehensive improvement and rational development of fragile mountain ecosystems [7]. In northwest Fujian, the ecological protection demonstration area is planned to be combined with constructing a farmland forest network and developing high-quality agriculture.

Furthermore, there are considerable factors related to the region's ecological aspects of land and water. Water resources should be used sparingly, with emphasis on ensuring the safety of centralised drinking water source protection areas and building urban reserve water sources. The government plans to strengthen the monitoring of water quality in drinking water sources, establish and improve an early warning system for drinking water source safety, and a monthly water quality report system to ensure drinking water safety. Comprehensive river basin management is also critical from social and environmental perspectives. The government aims to speed up the establishment of river basin environmental protection planning and rational definition of water regional environmental capacity. They aim to pay equal attention to protection and management, improve water quality in key river basins, and ensure that the water quality at the boundary of river basins' administrative regions reaches the standard [7]. The rational use of land resources would highly depend on the rational allocation of construction land.

According to laws and regulations and relevant planning, based on the quantity, quality, and ecological balance of cultivated land in the province, the layout of agricultural land should be adjusted scientifically and rationally. They should also rationally arrange and supply construction land in key areas. In this regard, the government aims to strengthen land consolidation, reclamation, and development, strictly protect cultivated land, and strictly observe the red line of 19.1 million mu of cultivated land [7]. The proposed plans will strengthen the protection of bare farmland to ensure that the amount of basic farmland in the province is not reduced, the use is not changed, and the quality is improved.

The other aspect is to invigorate idle land and the stock of construction land actively. The government aims to intensify the consolidation of construction land, reorganise abandoned land and "hollow villages" of industrial and mining enterprises, clean up and integrate industrial parks, implement management of industrial land investment intensity, and improve land use efficiency. These strategies adhere to economic and intensive use of land, coordinate the layout of construction, and concentrate industrial and living land according to functional areas to guide the transformation of urban construction land growth from extensive to intensive. The approaches would require spatial strategies for physical infrastructural development to support society. The next steps (in the near future) aim to clarify the plan for the construction of land supply. In order to prevent the disorderly expansion of urban land scale, the plan is to formulate the urban construction planning in the near future, and determine the scale of construction land and the time sequence of development.

9.4.2 *Sustainable Development Plan from the Economic Perspective*

Under the economic perspective, four key development plans are proposed for the GFZ city cluster. These are discussed briefly here:

(1) ***Energy infrastructure development strategy***
The government plans to fully use the excellent port conditions, actively use international and domestic resources, strengthen energy security, and optimise the energy structure. The foreseeing development is proposed to be combined with constructing coastal coal ports with a reasonable layout of coastal large-scale coal power. It is envisaged to consider a timely promotion of supercritical infrastructural development with an ultra-supercritical thermal power construction. The government aims to further deepen and improve preliminary work on nuclear power plant sites and speed up the construction of nuclear power projects in Ningde and Fuqing [7]. The government also plans to actively develop new energy sources such as wind and tide. These are set to further improve power grid construction, strengthen networking with East China Power Grid, and accelerate networking with the Southern Grid. The secondary plan is to build oil reserve bases and coal transfer bases in a comprehensive way. In doing so, the regional development shall strengthen cooperative exploration and joint development of oil and gas resources in the Taiwan Strait[7]. In combination with the layout of LNG receiving stations, the plan is to speed up the construction of gas transmission networks and gradually increase the civil gas coverage.

(2) ***Power development priorities***
In line with local conditions, the layout of coal power, nuclear power, gas power, wind power, and other power sources should be rationally optimised [7]. The construction of gas power plants includes the recent construction of the Putian gas power plant, Jinjiang gas power station, and Xiamen gas plant. The government plans to speed up the development of nuclear power, continue the construction of the Ningde and Fuqing nuclear power stations, and advance preliminary work on backup sites for nuclear power. The other power development proposals would strictly control small hydropower development and speed up the construction of Xianyou and other pumped storage power stations [7]. In doing so, the proposed plans encourage the development of waste-to-energy incineration, develop and build large-scale onshore wind power projects in an orderly manner, and actively promote offshore wind power demonstration projects. In the coming years and until the carbon peak in 2030, the government will arrange for the development of coal-fired power generation in a scientific manner, focusing on developing high-efficiency coal-fired generating units with a capacity of 600,000 kilowatts or above in supercritical and ultra-supercritical units. This approach leads to building large-scale coal-fired electricity generation in the regional ports.

(3) ***Oil and gas development priorities***
The proposed regional development plans aim to expand foreign cooperation in oil and natural gas, improve the production capacity of Fujian's petrochemical industry, build a pipeline from Fuzhou to Zhangzhou (the border between Fujian and Guangdong Province), and carry out preliminary research on the Zhangzhou-Longyan-Ganzhou pipeline [7]. The Fujian LNG pipeline network will be connected with the three networks of Zhejiang and Guangdong, extending the network to some parts of Jiangxi. It will comprehensively cover the economic zone on the west bank of the Taiwan Strait and realise the connection of the Fujian pipeline network from Ganzhou and Nanchang to the second line of west–east gas transmission, forming a "multi-source and multi-directional" gas supply pattern in the province [36]. The construction of Ningde and Xiuyu LNG receiving stations accelerates the province's "two vertical and two horizontal" gas transmission trunk network and branch pipeline network construction. It also promotes the formation of pipeline network patterns [45]. In this regard, the government can create conditions for actively carrying out preliminary work such as oil and gas exploration in the Taiwan Strait. Promote the establishment of oil and gas reserves.

(4) ***Renewable energy development priorities***
The recent focus on the development of wind power and biomass power generation takes into account solar energy, tidal energy, geothermal energy, rural methane, and other types of development and utilisation [7]. According to local conditions, the planned developments aim to speed up the promotion and development of rural biomass power generation, plant energy crops and plants, and develop a new rural energy industry. Aligned with this strategy, large-scale development of biomass energy resources is essential in achieving industrial development. In addition, the secondary plan is also to optimize the development of tourism resources, develop coastal tourism and strait tourism themes, and increase the proportion of tourism in the local industry.

9.5 Pathways to Carbon Neutrality in the Guangdong-Fujian-Zhejiang Coastal Region

In summary, the GFZ city cluster differs significantly from the earlier city clusters we have mapped in the book. The struggles for the GFZ city cluster are quite different from the earlier cases, as the focus is still on economic development. Thus, as a network corridor between medium-sized cities, the importance of economic growth and development seems to be very visible in this coastal region. A project example is the Fujian Three Gorges Maritime Wind Power Industrial Park, which aims to boost the carbon peak achievements:

> The Fujian Three Gorges Maritime Wind Power Industrial Park is a strategic cooperation project between Three Gorges group and Fujian Province. A staff member of the park told the reporter that the eastern region is short of onshore wind power compared with the northeast,

> north and northwest of China. The park thus makes use of its geographical advantages located in coastal areas and the location advantages close to the eastern load center to vigorously develop offshore wind power [46].

Other aspects to consider are the regional focus on development zones and meeting the carbon peak targets of 2030. For instance, two aspects of 'coefficient of carbon emission' and 'ecological support coefficient of carbon absorption' [53] are highlighted as critical characteristics and strategies of carbon neutrality plan at the provincial level (in Fujian). Therefore, industrial development and its integration into the extended urban and regional plans are expected to play a major part in defining the GFZ's carbon emission reduction targets. In reality, the region has to focus on equity and efficiency factors [5, 47] to ensure carbon neutrality targets are met while urbanisation and zone development are still ongoing.

References

1. Bund S (2016) 12 Regenerating Urban Waterfronts in China. Waterfronts revisited: European ports in a historic and global perspective, p 163
2. Chen X (2007) A tale of two regions in China: rapid economic development and slow industrial upgrading in the Pearl River and the Yangtze River Deltas. Int J Comp Sociol 48(2–3):167–201
3. Cheshmehzangi A (2014) The urban and urbanism: China's new urbanization and approaches towards comprehensive development. Int J Interdiscip Environ Stud 8(3–4):1–12
4. Cheshmehzangi A (2016) China's New-Type Urbanisation Plan (NUP) and the foreseeing challenges for decarbonization of cities: a review. Energy Procedia 104:146–152
5. Cheshmehzangi A, Chen H (2021) Key suggestions and steps ahead for China's carbon neutrality plan. In: China's sustainability transitions. Springer, Singapore, pp 189–206
6. Cheshmehzangi A, Dawodu A, Sharifi A (2021) Resilient city development: Best practices through the effectiveness of Resilience Index System. In: Cheshmehzangi A, Dawodu A, Sharifi A (eds) Sustainable urbanism in China. Routledge, New York
7. China City News (2021) Guangdong, Fujian and Zhejiang coastal urban agglomeration replace the West Bank City of the strait, change only the name? What does it mean?中国城市报(2021).<粤闽浙沿海城市群取代海峡西岸城市群, 变的只是名称?有何深意?>. Available from: https://www.163.com/dy/article/G5Q16VBR0530RTB0.html
8. China Industry Development and Research Network (2015) Analysis of current situation and future development planning of urban agglomeration on the West Bank of the straits <海峡西岸城市群发展现状及未来发展规划分析 - 宏观经济 - 中国产业发展研究网>. Available from: http://www.chinaidr.com/news/2015-01/39678.html
9. Choi CC (2010) Stepping out? Women in the Chaoshan Emigrant. Merchants' Daughters: Women, Commer, Reg Cult South China 1:105
10. Delgado M, Porter ME, Stern S (2014) Clusters, convergence, and economic performance. Res Policy 43(10):1785–1799
11. Dong Q, Yi P, Li W, Wang L (2022) Evaluation of city sustainability using the HGRW method: A case study of urban agglomeration on the West Side of the Straits, China. J Clean Prod 358:132008
12. Forster K (1990) The Wenzhou model for economic development: impressions. China Inf 5(3):53–64
13. Fujian Provincial People's Government (2010) Development Planning of Urban Agglomerations on the West Bank of the Straits. <海峡西岸城市群发展规划>. Available from: http://www.cre.org.cn/zl/xlwjlz/csqgh/15385.html

14. Gan GY, Su JY (2021) A new reclassification in international container transport market based on the impact of “One Belt And One Road” initiative. J Mar Sci Technol 29(1):2
15. Gao K, Yuan Y (2021) The effect of innovation-driven development on pollution reduction: empirical evidence from a quasi-natural experiment in China. Technol Forecast Soc Chang 172:121047
16. Head K, Ries J, Sun X, Hong J (2015) The legacy of nineteenth century treaties on the current trade of Chinese cities. Asia-Pac J Account Econ 22(3):251–270
17. Hsu JY (2006) The dynamic firm–territory nexus of Taiwanese informatics industry investments in China. Growth Chang 37(2):230–254
18. Huang Z, Zhang X, Yunwei ZHU (2008) The role of clustering in rural industrialization: a case study of the footwear industry in Wenzhou. China Econ Rev 19(3):409–420
19. Jiang W, Yu Y, Zhao T, Yao C, Tang K, Liu J (2018) A summary of innovation and reform of economic development pattern in New Era. In: 2018 8th international conference on social science and education research (SSER 2018). Atlantis Press, pp 463–471
20. Lai HC, Chiu YC, Leu HD (2005) Innovation capacity comparison of China’s information technology industrial clusters: the case of Shanghai, Kunshan, Shenzhen and Dongguan. Technol Anal Strat Manag 17(3):293–316
21. Li C (2020) The establishment and basic characteristics of western rules of travel to China. J Landsc Res 12(5)
22. Li X, Lin T, Zhang G, Xiao L, Zhao Q, Cui S (2011) Dynamic analysis of urban spatial expansion and its determinants in Xiamen Island. J Geog Sci 21(3):503–520
23. Li X, Zhang Q (2015) AHP-based resources and environment efficiency evaluation index system construction about the west side of Taiwan Straits. Ann Oper Res 228(1):97–111
24. Li Y, Sun Z, Mao L, Hu X, Chen B, Li Y (2022) Spatial variation and ecological risk assessment for heavy metals in marsh sediments in Fuzhou reach of the Min River, Southeast China. Mar Pollut Bull 180:113757
25. Li Z (2015) The two urban planning efforts of Shantou in the city hall era of the Republic of China. China City Plan Rev 24(4)
26. Lin S, Li Z (2019) City profile: Wenzhou-a model city of transitional China. Cities 95:102393
27. Lin X, Lu C, Song K, Su Y, Lei Y, Zhong L, Gao Y (2020) Analysis of coupling coordination variance between urbanization quality and eco-environment pressure: a case study of the west taiwan strait urban agglomeration, China. Sustainability 12(7):2643
28. Liu YL (1992) Reform from below: the private economy and local politics in the rural industrialization of Wenzhou. China Q 130:293–316
29. Liu AP (1992b) The” Wenzhou model” of development and China’s modernization. Asian Surv 696–711
30. Lu C, Li L, Lei Y, Ren C, Su Y, Huang Y, Chen Y, Lei S, Fu W (2020) Coupling coordination relationship between urban sprawl and urbanization quality in the West Taiwan Strait urban agglomeration, China: observation and analysis from DMSP/OLS nighttime light imagery and panel data. Remote Sens 12(19):3217
31. Luo F (2015) Merchant guilds, local autonomy, and municipal governance in a South China treaty port: Shantou, 1858–1939. Illinois State University
32. Ma Y, Xue F (2019) Deciphering the spatial structures of city networks in the economic zone of the West Side of the Taiwan Strait through the lens of functional and innovation networks. Sustainability 11(10):2975
33. Mark Talk History (2021). Five ports of trade in five cities, three are the world’s top ten ports, five are the top 100 ports in the world.马克聊历史(2021).<五口通商的五座城市,3座已是全球十大港口,5座都是全球百大港口>. Available from: https://www.360kuai.com/pc/9b05797376df92760?cota=3&kuai_so=1&sign=360_57c3bbd1&refer_scene=so_1
34. McKenney KI (1993) An assessment of China’s special economic zones. Industrial Coll of the Armed Forces Washington DC. Available from: https://www.apps.dtic.mil/sti/citations/ADA276611
35. Meng G, Zeng DZ (2019) Structural transformation through free trade zones: the case of Shanghai. Transnatl Corp J 26(2)

36. Offshore Energy (2021) PipeChina connects Yuedong LNG terminal to three cities, a report under business developments and projects. Available from: https://www.offshore-energy.biz/pipechina-connects-yuedong-lng-terminal-to-three-cities/
37. Parris K (2017) Local initiative and national reform: the Wenzhou model of development. In: Chinese Economic History Since 1949. Brill, pp. 1479–1502. Available from: https://brill.com/downloadpdf/book/9789004304987/B9789004304987-s063.pdf
38. QQ News (2021) From the urban agglomerations on the west bank of the Taiwan Strait to the coastal cities of Guangdong, Fujian and Zhejiang, the names are not the only changes <从海峡西岸城市群到粤闽浙沿海城市群, 变化的不仅仅是名称>. Available from: https://new.qq.com/rain/a/20210323A01PK400
39. Roberts PM (2012) Nanjing, Treaty of (August 1842). China at war: an encyclopedia, p 298
40. Sonobe T, Hu D, Otsuka K (2004) From inferior to superior products: an inquiry into the Wenzhou model of industrial development in China. J Comp Econ 32(3):542–563
41. Stern N, Xie C (2021) The 14th five-year plan: peaking in China's greenhouse gas emissions and paving the way to carbon neutrality. Grantham Research Institute on Climate Change and the Environment, Available at: https://www.lse.ac.uk/granthaminstitute/wp-content/uploads/2021/07/The-14th-Five-Year-Plan_Peaking-Chinas-greenhouse-gas-emissions-and-paving-the-way-to-carbon-neutrality.Pdf. Accessed 29
42. Stern N, Xie C (2022) China's new growth story: linking the 14th Five-Year Plan with the 2060 carbon neutrality pledge. J Chin Econ Bus Stud 1–21
43. Tan GL (2018) An Introduction to the culture and history of the Teochews in Singapore. World Scientific, Singapore
44. Tang L, Zhao Y, Yin K, Zhao J (2013) Xiamen. Cities 31:615–624
45. Teller Report (2021) The construction of the ningde covered bridge in Fujian is handed down from generation to generation: if the bridge is there, we are here. Available from: https://www.tellerreport.com/life/2021-07-21-the-construction-of-the-ningde-covered-bridge-in-fujian-is-handed-down-from-generation-to-generation--if-the-bridge-is-there--we-are-here.B1-OvLIHR_.html
46. The People's Government of Fujian Province (2022) A date with China: "Blue Engine" boosts achievement of carbon peak, neutrality in E China's Fujian. Available from: https://www.fujian.gov.cn/english/news/202208/t20220805_5969316.htm
47. Tian M, Hu YJ, Wang H, Li C (2022) Regional allowance allocation in China based on equity and efficiency towards achieving the carbon neutrality target: a composite indicator approach. J Clean Prod 342:130914
48. Wei YD, Li W, Wang C (2007) Restructuring industrial districts, scaling up regional development: a study of the Wenzhou model, China. Econ Geogr 83(4):421–444
49. Wu W (1999) City profile: Shanghai. Cities 16(3):207–216
50. Xiamen Real Estate Information (2021) Super planning introduced! "West Strait City Group" upgraded to "Guangdong, Fujian and Zhejiang city group". <超级规划出台! "海峡西岸城市群"升级为"粤闽浙城市群" >. Available from: https://www.sohu.com/a/455750290_404513
51. Xiamen University (2021) Challenges and Opportunities of Coastal Urban Agglomeration in Guangdong, Fujian and Zhejiang <新提法粤闽浙沿海城市群的挑战与机遇>. Available from: https://baijiahao.baidu.com/s?id=1694624570124465538
52. Xie Y (2020) Research on concentration force of goods in ports across the Taiwan strait. Master dissertation in International Transport and Logistics at The World Maritime University, Shanghai, China. Available from: https://commons.wmu.se/cgi/viewcontent.cgi?article=2601&context=all_dissertations
53. Xu Y, Guo N, Ru KL, Fan SL (2022) Characteristics and optimization strategies of territorial space zone in Fujian Province, China based on carbon neutrality. Ying Yong Sheng tai xue bao = J Appl Ecol 33(2):500–508
54. Yao S, Zhao R, Xu G, Zhang L (1991) Some problems on regional development in Xiamen City. Chin Geogra Sci 1(1):70–82
55. Ye C, Zhu J, Li S, Yang S, Chen M (2019) Assessment and analysis of regional economic collaborative development within an urban agglomeration: Yangtze River Delta as a case study. Habitat Int 83:20–29

56. Ye X, Wei YD (2005) Geospatial analysis of regional development in China: the case of Zhejiang Province and the Wenzhou model. Eurasian Geogr Econ 46(6):445–464
57. Yeh AG, Liu X, Xu J, Wu M (2020) The emergence and economic restructuring of two global super megacity-regions in China: comparing the Pearl River and Yangtze River Deltas. In: Handbook of Megacities and megacity-regions. Edward Elgar Publishing, pp 376–394
58. Zeping H (2020) The new trend of the great migration in China. <中国人口大迁移的新趋势>. Available from: https://www.sohu.com/a/416765981_467568
59. Zhan S (2018) BRICS summit and improvement of international fan urban environment in Xiamen. Meteorol Environ Res 9(3):109–113
60. Zhang X, Lin J (2011) Notice of Retraction: research on China southeast high speed railway's influence on the talented persons introduction of the economic zone of West Side of the Taiwan Straits. In: 2011 international conference on business management and electronic information, vol 4. IEEE, pp 123–128
61. Zhao S, Jiang Y, Wang S (2019) Innovation stages, knowledge spillover, and green economy development: moderating role of absorptive capacity and environmental regulation. Environ Sci Pollut Res 26(24):25312–25325
62. Zhou Q, van den Bosch CCK, Chen J, Zhang W, Dong J (2021) Identification of ecological networks and nodes in Fujian province based on green and blue corridors. Sci Rep 11(1):1–18
63. Zhou Y (2019) Report on the development of the Xiamen special economic zone. In: Annual report on the development of China's special economic zones. Springer, Singapore, pp 141–153
64. Zhou Y, Sun Y, Wei YD, Lin GC (2011) De-centering 'spatial fix'—patterns of territorialization and regional technological dynamism of ICT hubs in China. J Econ Geogr 11(1):119–150
65. Zhu W (2018) Research on the design of Xiamen's urban visual identity system. In: 2018 2nd international conference on economic development and education management (ICEDEM 2018). Atlantis Press, pp 222–225
66. Zhu W, Xu L, Tang L, Xiang X (2019) Eco-efficiency of the Western Taiwan straits economic zone: an evaluation based on a novel eco-efficiency model and empirical analysis of influencing factors. J Clean Prod 234:638–652
67. Zhuang Z, Fu S, Lan S, Yu H, Yang C, Huang GQ (2021) Research on economic benefits of multi-city logistics development based on data-driven analysis. Adv Eng Inform 49:101322

Chapter 10
Guanzhong Plain City Cluster: A Rising Inland City Cluster with a Complete Industrial System

10.1 A Brief Introduction to the Guanzhong Plain City Cluster

The planning scope of the Guanzhong Plain (GP) city cluster includes Shaanxi Province and its five cities of Xi'an, Baoji, Xianyang, Tongchuan, and Weinan. It also includes Yangling Agricultural High-tech Industrial Demonstration Zone and Shangzhou District, Luonan County, Danfeng County, and Zhashui County. The GP city cluster also includes Yuncheng City of Shanxi Province (except for Pinglu County and Yuanqu County), Yaodu District of Linfen City, Houma City, Xiangfen County, Huozhou City, Quwo County, Yicheng County, Hongdong County, and Fushan County. Cities in the Gansu Province include Tianshui City, Pingliang City, Kongtong District, Huating County, Jingchuan, Chongxin, Lingtai, and Qingyang City. The GP city cluster covers a considerable area of approximately 107,100 square kilometers. In 2021, the region's estimated population was 42.34 million, and its GDP was at 2,193.32 billion, equivalent to a per capita GDP of 51,800 yuan.

The GP city cluster is relatively significant in China's inland regions. This urban agglomeration, centered in Xi'an, is an important birthplace of Chinese civilization and the starting point of the ancient Silk Road [11]. It bears the historical glory and heavy memory of the Chinese nation. The GP city cluster has a good foundation and great potential for development. It has a unique strategic position in the overall situation of national modernisation and all-around opening up [22]. It is located in the center of China's inland regions and is an important fulcrum of the 'Eurasian Continental Bridge' [42]. It is also the second largest city cluster in China's western regions. In 2017, the comprehensive economic strength of the city cluster was only second to the Chengdu-Chongqing city cluster, and ranked second in the inland regions [8]. The north–south passageway through the western region and the New Eurasian Continental Bridge meet here [1, 26]. Moreover, the critical high-speed railway network and highway network centered in Xi'an [39] accelerate the gathering of population and economic factors [28] in the GP city cluster and Xi'an Metropolitan Area.

A. Cheshmehzangi and T. Tang, *China's City Cluster Development in the Race to Carbon Neutrality*, Urban Sustainability, https://doi.org/10.1007/978-981-19-7673-5_10

The Guanzhong Plain Urban Agglomeration aims to build a state-level city cluster with international influence and a new highland of inland reform and opening up [23]. It plans to promote the construction of the core city of Xi'an into a national central city. In doing so, the plans are to enhance further the comprehensive carrying capacity of important nodes such as Baoji, Tongchuan, Weinan, Yangling, Shangluo, Yuncheng, Linfen, Tianshui, Pingliang, and Qingyan. It also aims to moderately expand the urban population and improve the comprehensive service function.

10.1.1 The Population Structure of the Guanzhong Plain City Cluster

Many factors affect population migration, such as nature, geography, economy, politics, society, culture, etc. The relevant theories include Ravenstein migration law, push–pull theory, dual structure theory of development economics, etc. [19, 24]. From the long-term and fundamental point of view, the key to determining the concentration of a region's population is the size of the region's economy and the per capita income gap between the city and the rest of the country. In theory, under the condition of complete market competition and individual homogeneity [37], higher per capita income will continue to attract net immigration from outside the region until the per capita incomes in the region are equal to those in the other areas. In short, people follow industry [34, 36].

Since 2010, with the industrial transformation and upgrading in coastal areas, the industrial undertaking in the central and western regions, and the aging of the older generation of migrant workers, some of the population has gradually returned to the Midwest regions [27]. According to the primary data of the seventh national census in Shaanxi Province, Xi'an has increased by 4.4851 million people compared to 2010 [22]. The number of permanent residents in Xian reached a staggering 12.95 million, accounting for nearly one-third of the province's overall population. The population increment ranked third in China and first in the north. In this regard, Xi'an has become the "third city" in the north and ranks among the top ten in the country [41]. Xi'an's population surges behind, not only in the northwest bridgehead position but also in Shaanxi and to the west of Inner Mongolia, southwest Shanxi, Henan, and other regional areas. Within the GP city cluster, it is evident that Xi'an has attracted most of the talents, young workers, and professionals.

10.2 A Historical Analysis of the Guanzhong Plain City Cluster

The Guanzhong Plain is an important birthplace of Chinese civilization and the starting point of the ancient Silk Road [14, 31]. It has a unique strategic position in the overall situation of China'sChina's modernisation and all-around opening up. The regional governments continue to foster and develop the urban agglomeration in the Guanzhong Plain, giving play to its geographical advantages as a link between the east, the west, and the south. Promoting economic growth and expanding market space from the east to the west and from the south to the north will help guide and support the development and opening up of the northwest region, advance the development of the western region, and further advance the Belt and Road Initiative (BRI). Xi'an was the capital of several dynasties in Chinese history, and it was also the hub of the central dynasty controlling the northwest region and connecting the central plains with the northwest and southwest regions [25, 29, 35]. After the socialist period (with Chinese characteristics), the city entered a new era [41]. It is a significant academic and practical issue to adjust and reform the administrative division of the Guanzhong region with Xi'an as the national central city.

Through the establishment of administrative districts in the Guanzhong region, we note the growing importance of city cluster development for its long-term plans. By taking Xi'an as its center, the GP city cluster demonstrates the important characteristics of forming multiple phases. It is concluded that merging Xianyang and Tongchuan into Xi'an can reproduce the classical administrative division structure of the Western Han Dynasty and the Eastern Han and Sui Dynasties. QinDu District and Weicheng District, formerly under the jurisdiction of Xianyang City, were listed as the municipal districts of Xi'an, and Weicheng District was renamed "Xian Yang District" [41]. Huangling County was included in the city of Xi'an, increasing the influence of Pure Brightness of the Yellow Emperor Mausoleum sacrifice. Finally, the Fuping County of Weinan City was transferred into Xi'an City. In addition, Weiyuan County, Longxi County of Dingxi City, Gansu Province, the source of the Weihe River, the Jingyuan City of Guyuan, and Ningxia Hui Autonomous Region are also included in the GP city cluster. In this way, the city cluster, led by Xi'an, a state-level central city, is formed as the economic up of the north-western regions of China [41]. Xi'an, a national central city, is expected to be built into an economic center in western China, a highland for foreign exchanges under the BRI, a cultural highland of the Silk Road, an inland opening highland, and a comprehensive transportation hub.

10.3 An Overview of the Industrial Structure in the Guanzhong Plain City Cluster

The Guanzhong Plain city cluster has a complete industrial system and a high degree of industrial concentration. It is an essential base for the equipment manufacturing industry, the high-tech industry, and the national defense science and technology industry [41]. Aviation, aerospace, new materials, a new generation of information technology, and other strategic emerging industries are developing rapidly. There is a lot of emphasis on developing culture, tourism, logistics, finance, and other modern service industries, which are proliferating. In recent years, in particular, we note that the industrial structure is moving towards the high-end. The GP city cluster is the only free trade zone and a batch of national industrial parks in northwest China, which provides an important platform and carrier for the development of the modern industry. As the State Council executive meeting on the 31st of August 2018 highlighted, the city cluster "*must continue to deepen the reform of government functions, to foster a world-class, market-oriented business environment government by a sound legal framework*" [22]. As the starting point of the "Silk Road Economic Belt", it is almost inevitable to have a complete industrial system that could help the regional economic growth.

A recent study by [10] highlights the coupling coordinated relationship between China's new-type urbanisation plan (NUP) [4, 5] and eco-development in the Guanzhong region. In this study, issues of industrial agglomeration effect are highlighted through which a few points are discussed, such as the "*industrial upgrading effect, scale economy effect, and comparative advantage effect*" [10]. It is important to note the adjustment of industrial structure is correlated with the population increase, changes in lifestyles, land, energy-use patterns, and other variables. Thus, despite the attempts at transitional developments in the GP city cluster, the region's "*industrial development model of high energy consumption, high pollution, high emissions, and low returns has not been fundamentally changed*" (ibid). In fact, the regional urbanisation trend is blended with industrialisation and relevant economic structure, which suggests the unique characteristics of the GP city cluster. As the foundation of its urbanisation pattern, industrialisation in the GP city cluster plays a major part in accelerating regional development and supporting a better "*flow of labour, capital, and land*" (ibid). Yet, structural upgrading and industrial transformation are highlighted as significant obstacles for the GP city cluster, indicating the inequalities between different areas in the region [2, 9, 18, 40]. The industrial structure adjustments are yet to be considered, and the region's industrial completeness is expected to make the forthcoming transitions to be complicated:

> At the same time, the economic structure within the urban agglomeration is converging, the industrial level is low, and traditional industrial urbanization with high input, high consumption, and high emissions is the mainstay. In addition, the fragile background ecology in the northwest has increased the bearing pressure on resources and the environment, resulting in a relatively high proportion of low-coupling and coordination areas in the Guanzhong Plain urban agglomeration. The unsustainability of this development model needs to be alleviated

urgently. This is basically consistent with previous research conclusions in other regions [10].

Moreover, much of the industrial structure is linked to the pattern change and transitional development of urban economic growth [20], with requires upgrading and readjusting. Hence, a green transformation process is required to guide the cross-regional development patterns [10], particularly through the strengthening of scientific and technological innovation.

10.4 The Guanzhong City Cluster's Sustainability Agenda: Plans and Perspectives

This section briefly discusses the GP city cluster's sustainability agenda. We highlight key aspects of the regional sustainable development plan from social, environmental, and economic perspectives. The following aspects are extracted from relevant government plans, documents, and reports.

10.4.1 Sustainable Development Plan from the Social Perspective

The government plans to strengthen innovation in critical areas of competitive industries and extend the industrial chain. The primary focus is on energy and chemical industries, high-end equipment manufacturing, aviation, aerospace, new generation information technology, and new materials. Relying on industrialisation platforms such as the 'Innovation Port in Western China and the 'National Engineering Research Center' [38], the emphasis is on innovation and professional development [41]. At least 15 major industrial innovation projects are proposed to be implemented in the coming years. Efforts should be made to break through the technical bottlenecks of missing links, weak links, and extended links in the industrial chain. Thus, such development plans should promote the research and development of key industrial technologies and the application of advanced technological achievements and push industries to climb from the low-end to the high-end of the value chain.

The other key aspect is to rely on the advantages of the innovation chain. This approach helps cultivate emerging industrial chains [41]. By relying on national innovation platforms, such as additive manufacturing, molecular medicine transformation, ceramic matrix composites, and superconducting materials, the government aims to accelerate the local transformation of high-tech and new technological achievements. It also aims to foster industrial clusters such as 3D printing, biomedicine, new materials, integrated circuits, and artificial intelligence (AI) to build strategic emerging industries.

Lastly, the plan is to improve regional innovation networks and optimise innovation in the ecological environment [3]. Xi'an High-tech Zone, as a new model of development in the region, plays a leading role in the innovation of the 'National Independent Innovation Demonstration Zone', and promotes the innovative development of Xian New Area and Baoji, Xianyang, Weinan, Yangling, and Tianshui development zones [41]. In this regard, the government supports the establishment of demonstration bases for entrepreneurship and innovation, fosters and develops a system of financial intermediation services in science and technology, and supports organisations that provide intermediary financial services in the fields of intellectual property evaluation, technology transfer, patent agency, and credit rating. Thus, the government has ongoing projects to build a policy environment, institutional environment, and service system conducive to mass entrepreneurship and innovation.

10.4.2 Sustainable Development Plan from the Environmental Perspective

The GP city cluster's ecosystem is relatively fragile, and resources and environmental constraints are intensified. Per capita water resources are less than one-third of the national average level, and the spatial distribution is uneven. Water pollution is serious in some parts of the Weihe and Fenhe river basins, and the problem of groundwater over-exploitation is prominent [13]. The air pollution situation is severe [41]. Consequently, we see that heavy pollution weather occurs frequently, and the environmental capacity is close to the limit.

A major regional sustainable development plan focuses on improving water resources security capacity. The government aims to strengthen the protection of drinking water sources and speed up projects to divert key water sources such as the Hanji-Wei and Jinghe Dongzhuang reservoirs. They plan to deepen preliminary work on the Bailong River diversion project and the Malian River Water Control Project in Qingyang and strengthen the construction of emergency backup water sources [41]. This approach helps build a water supply guarantee system that is interconnected and where multiple water sources are provided in this capacity [33]. The preliminary plans are to speed up the implementation of water-saving improvements in large and medium-sized irrigation districts such as the Baoji Gorge and the Jinghui Canal, improve the recycling rate of industrial water use, promote water conservation in daily life, and promote the sustainable use of water resources [41]. In doing so, the ongoing projects accelerate the construction of water-saving cities, focusing on cities above the prefectural level that lack water. The imposed strategies help regulate rivers and lakes systematically, maintain the ecological base of river courses, repair the ecological environment of lakes and wetlands [16], strengthen flood storage and utilisation, and realise the goal of "*water moistening, water beautifying, and water prospering*".

Lastly, another major sustainable development plan is to improve the water resources management system. The government aims to fully implement the strictest water resources management system, establish a reasonable water pricing mechanism, and strictly control the total amount of water used. The ongoing projects are focused on improving the coordination and management mechanism for the joint protection, development, and utilisation of water resources. They also help speed up the establishment of regional water resources information platforms and realise the sharing of basic information on regional waters and the networking of monitoring and monitoring [41]. In this regard, the government aims to improve the modern management level of water consumption measurement and supervision, hydrological measurement and forecasting, water quantity and quality monitoring, and water resources regulation [32]. The red line of functional water zones to limit the intake of pollutants regulate the supervision and management of sewage outlets entering rivers and lakes [12]. It also further advances the construction of water ecological civilization, implement the river chief system in all areas, and implement the lake chief system on lakes, so as to bring pollution systems under control in the river basins.

10.4.3 Sustainable Development Plan from the Economic Perspective

The establishment of an urban agglomeration ecological protection coordination mechanism is central to the economic growth plans at the regional level [15]. The sustainable development plan of the GP city cluster is to promote the integrated construction of ecological barriers in urban areas [30], with emphasis on the integrated protection and management of the Qinling Mountains, the Yellow River, and the Weihe River. It is also proposed to increase input in biodiversity conservation, water conservation, and the management of mountainous areas along the Huanghe River's coast and jointly create an eco-economic belt of the river [41]. At the city cluster development level, the development of the enclave economy and the sharing of public resources are encouraged to establish a horizontal compensation mechanism between the ecological benefit areas and the ecological protection areas. The intention is, therefore, to develop a mechanism for sharing costs and benefits [6]. In this regard, the main economic development projects are to establish cost-sharing mechanisms for major trans-regional infrastructure, public services, and ecological environment construction projects. Thus, the government and related state-led organisations promote the co-construction and sharing of transportation, communications, water conservancy, and other infrastructure and implement a smart "one card" for public transportation. The government has also proposed to establish a unified platform for sharing science and technology resources, establish a coordination mechanism for science and innovation, and promote the integration of science, technology, and innovation policies.

The other economic growth is to strengthen docking with the eastern coastal city group, which is vital for promoting economic prosperity at the regional level. In this regard, the government gives prominence to cooperation in the east with the Beijing-Tianjin-Hebei, Yangtze River Delta, and Guangdong-Hong Kong-Macao regions, relying on land bridges and sea passageways. This strategic plan would vigorously strengthen cooperation with the eastern regions in technological research and development, ecological and environmental protection, public services, investment, and finance. In the coming years, the government plans to strengthen cooperation with Tianjin Port, Qingdao Port, and Lianyungang Port in eastern China. The aim is to promote the connection of sea, land, and airports, realise linkage between land and sea, and speed up the integration of customs clearance. In addition, the plan is to expand opening up and cooperation with Hong Kong, Macao, and Taiwan regions.

More importantly, the GP city cluster plans to strengthen cooperation with surrounding urban agglomerations. This is more of the win–win cooperation strategies that could vigorously promote exchanges and cooperation to the west, south, and north and strengthen coordination and interaction with the urban agglomerations of the northwest region. The government explores the establishment of cooperation mechanisms with the urban agglomeration on the middle reaches of the Yangtze River and the Chengdu-Chongqing urban agglomeration and promotes the coordinated development of the urban agglomerations in the central plains, central Shanxi, and Hohhot, Baotou, Hubei, and Yu regions. The strategic geographical location of the GP city cluster enables it to be well-connected to its neighbouring regions. To achieve mutually beneficial cooperation with provinces (regions) along the 'Silk Road Economic Belt', the primary economic development focus is on strategic planning, cultural tourism, infrastructure development, and industrial development. More importantly, the overarching plan is to promote the development of underdeveloped areas in the periphery. Thus, the government aims to strengthen cooperation and exchanges with contiguous poverty-stricken areas such as Qinba Mountain, Liupan Mountain, and Luliangshan Mountain, as well as the old revolutionary base areas of Shaanxi-Gansu-Ningxia and Sichuan-Shan. This strategy is the backbone of the GP city cluster's pathways to sustainable development, which require further efforts in strengthening connectivity between poverty alleviation and development, infrastructure, public services, and featured industries. Thus, the government has to increase the role of 34 sectors in promoting poverty alleviation and jointly speed up the pace of poverty alleviation. In doing so, the complete industrial system could be promoted and, at the same time, advantageous for the regional economic development.

10.5 Pathways to Carbon Neutrality in the Guanzhong Plain City Cluster

In the GP city cluster, rural–urban relations play a significant part as they help to promote renewable energy capacity and smart grid systems [7], and optimise the national grid. In this regard, relevant policies are suggested to support energy conservation and emission reductions, to "*strictly enforce environmental protection standards, and eliminate backward production capacity, so as to promote sustainable development of the Guanzhong Plain urban agglomeration*" [10]. The overall plan for the GP city cluster depends on its complex—yet complete—industrial system. The transformational industrial restructuring and necessary readjustments are the backbones of the GP's carbon neutrality plan. As a leading region for smart grid energy systems [6], urban–rural development strategies are important to achieving the carbon peak targets of 2030 and the carbon neutral plan of 2060.

In the coming years, the GP city cluster should focus on identifying and reducing air pollution, particularly from high-polluting industries. The complexity remains for the diverse range of sectors, various emission factors [21], characteristics of industries, etc. [17]. Thus, the development of legislation, standards, and policies will not be straightforward. In line with China's road to carbon neutrality, the GP's focus on industrial restructuring, relocation, and readjustments should continue until at least 2030. After that, and once the first steps are achieved successfully, the city cluster is expected to be in a better position in terms of its infrastructural capacity, grid system, and integrated strategies for carbon emission reductions. Thus, from 2030 onwards, the focus could be on extensive social and environmental aspects only if the infrastructural and communications links/connections are made, the scientific plans are adopted, and technological advancements have strengthened the institutional systems of the region by then.

References

1. Bai Y, Wang S (2021) Spirit of the silk road: Chinese trade and investment throughout the Eurasian Corridor. Palgrave Macmillan, Singapore
2. Bie X, Sun T, Sun R, Wang J, Zhang T, Chen L (2018) Urban expansion and its spillover effects on economic growth in the Beijing-Tianjin-Hebei urban agglomeration. Acta Ecol Sin 38(12):185–194. https://doi.org/10.5846/stxb201801310276
3. Chen J, Wang S, Zou Y (2022) Construction of an ecological security pattern based on ecosystem sensitivity and the importance of ecological services: a case study of the Guanzhong Plain urban agglomeration, China. Ecol Indic 136:108688. https://doi.org/10.1016/j.ecolind.2022.108688
4. Cheshmehzangi A (2014) The urban and urbanism: China's new urbanization and approaches towards comprehensive development. Int J Interdiscip Environ Stud 8(3–4):1–12
5. Cheshmehzangi A (2016) China's New-Type Urbanisation Plan (NUP) and the foreseeing challenges for decarbonization of cities: a review. Energy Procedia 104:146–152

6. Cheshmehzangi A, Chen H (2021) Lessons and paradigms to meet the 2030 targets and the 2060 plan. In: China's sustainability transitions: low carbon and climate-resilient plan for carbon neutral 2060. Springer, Singapore, pp 177–187
7. Cheshmehzangi A, Dawodu A, Sharifi A (2021) Smart grid energy systems: an integrated urban innovation to optimise energy use. In: Sustainable urbanism in China. Routledge, New York, pp 267–292
8. Chu YW (2020) China's new urbanization plan: progress and structural constraints. Cities 103:102736
9. Cui X, Fang C, Liu H, Liu X (2019) Assessing sustainability of urbanization by a coordinated development index for an urbanization-resources-environment complex system: a case study of Jing-Jin-Ji Region, China. Ecol Ind 96:383–391. https://doi.org/10.1016/j.ecolind.2018.09.009
10. Dong L, Shang J, Ali R, Rehman RU (2021) The coupling coordinated relationship between new-type urbanization, eco-environment and its driving mechanism: a case of Guanzhong, China. Frontiers in Environmental Science, Section: Environmental Informatics and Remote Sensing. https://doi.org/10.3389/fenvs.2021.638891
11. Frankopan P (2015) The silk roads: anew history of the world. Bloomsbury Publishing, London and New York
12. Gao C, Wu Q, Dyck M, Lv J, He H (2022) Greenhouse area detection in Guanzhong Plain, Shaanxi, China: spatio-temporal change and suitability classification. Available from: http://lib.cqvip.com/Qikan/Article/Detail?id=7107691021
13. Huo P, Li H, Huang X, Ma X, Liu L, Ji W, Liu Y, Gao P (2022) Dissolved greenhouse gas emissions from agricultural groundwater irrigation in the Guanzhong Basin of China. Environ Pollut 209:119714. https://doi.org/10.1016/j.envpol.2022.119714
14. Jiang W, Zou M (2021) Innovative design of campus green spaces in the context of the" One Belt, One Road" and the digital use of Silk Road elements. In: E3S web of conferences, vol 236. EDP Sciences, p 05031
15. Ma F, Wang Z, Sun Q, Yuen KF, Zhang Y, Xue H, Xhao S (2020) Spatial-temporal evolution of urban resilience and its influencing factors: evidence from the Guanzhong plain urban agglomeration. Sustainabilitym 12(7):2593. https://doi.org/10.3390/su12072593
16. National Development and Reform Commission (2019) Notice of the National Development and Reform Commission on printing and distributing the pilot scheme of ecological comprehensive compensation, Governmental Report, Beijing, China
17. Niu X, Cao J, Shen Z, Hang Ho SS, Tie X, Zhao S, Xu H, Zhang T, Huang R (2016) PM2.5 from the Guanzhong plain: chemical composition and implications for emission reductions. Atmos Environ 147:458–469. https://doi.org/10.1016/j.atmosenv.2016.10.029
18. Pan JH, Feng YY (2017) Estimating potential ecological carrying capacity in Gansu Province. Chin J Ecol 36(3):226–234. https://doi.org/10.13292/j.1000-4890.201703.024
19. Qian W (1996) Rural-urban migration and its impact on economic development. Avebury, Brookfield, USA
20. Ren W, Zhao J, Ma X (2022) Analysis of the spatial characteristics of inhalable particulate matter concentrations under the influence of a three-dimensional landscape pattern in Xi'an, China. Sustain Cities Soc 81:103841. https://doi.org/10.1016/j.scs.2022.103841
21. Sun J, Shen Z, Zhang L, Zhang Y, Zhang T, Lei Y, Niu X, Zhang Q, Dang W, Han W, Cao J, Xu H, Liu P, Li X (2019) Volatile organic compounds emissions from traditional and clean domestic heating appliances in Guanzhong Plain, China: emission factors, source profiles, and effects on regional air quality. Environ Int 133, Part B:105252. https://doi.org/10.1016/j.envint.2019.105252
22. The State Council (2018) Guanzhong plain urban cluster to embrace unified development plan. Available from: https://english.www.gov.cn/policies/latest_releases/2018/01/15/content_281476014917572.htm
23. Wan J, Yan J, Wang X, Liu Z, Wang H, Wang T (2019) Spatial-temporal pattern and its influencing factors on urban tourism competitiveness in City agglomerations across the Guanzhong plain. Sustainability 11(23):6743

24. Wang C (2021) How Migrants' Settlement Intention Is Shaped in China: Patterns, Determinants and Motivation (Doctoral dissertation, The Chinese University of Hong Kong, Hong Kong
25. Wang C, Ducruet C, Wang W (2015) Evolution, accessibility and dynamics of road networks in China from 1600 BC to 1900 AD. J Geog Sci 25(4):451–484
26. Wang JA, Liang S, Shi P (2022) China's opening up and trade. In: The geography of contemporary China. Springer, Cham, pp 337–376
27. Wang M, Chen Z, Zhang P, Tong L, Ma Y (eds) (2014) Old industrial cities seeking new road of industrialization: models of revitalizing Northeast China. World Scientific, Singapore
28. Wang X, Tomaney J (2019) Zhengzhou-political economy of an emerging Chinese megacity. Cities 84:104–111
29. Wang X, Zhao J (2012) The cultural exchange between Sino-Western: silk trade in Han Dynasty. Asian Cult Hist 4(1):13
30. Wei Z, Lei L, Lin L (2009) Study on sustainable development level of the central plain city cluster based on SOM neural network. Available from: http://en.cnki.com.cn/Article_en/CJFDTOTAL-HNKX200912035.htm
31. Xie B, Zhu X, Grydehøj A (2020) Perceiving the silk road archipelago: archipelagic relations within the ancient and 21 st-century maritime silk road. Isl Stud J 15(2)
32. Xu D, Xu L, Zhou Q (2020) The green and circular development strategy of the Qinba Mountain Area (Phase II). Strat Study Chin Acad Eng 22(1). Available from: https://doi.org/10.15302/j-sscae-2020.01.001
33. Yang Q, Zhang D (2021) The influence of agricultural industrial policy on non-grain production of cultivated land: a case study of the "one village, one product" strategy implemented in Guanzhong Plain of China. Land Use Policy 108:105579. https://doi.org/10.1016/j.landusepol.2021.105579
34. Yeung YM, Yeung YM, Shen J (eds) (2004) Developing china's west: a critical path to balanced national development. Chinese University Press
35. Yimin H, Lingwei W (2017) Ancient Chengdu and the Silk Road. In: Symposium on the land of abundance and the silk road. No. 5, Article 7, Available from: https://css.researchcommons.org/cgi/viewcontent.cgi?article=1082&context=journal
36. Yu J, Zhou K, Yang S (2019) Land use efficiency and influencing factors of urban agglomerations in China. Land Use Policy 88:104143
37. Zhang S, Zhang F, Wang C, Wang Z (2020) Assessing the resilience of the belt and road countries and its spatial heterogeneity: a comprehensive approach. PLoS ONE 15(9):e0238475
38. Zhao H, Niu Z, Feng X (2021) Factors influencing improvements in air quality in Guanzhong cities of China, and variations therein for 2014–2020. Urban Clim 38:100877. https://doi.org/10.1016/j.uclim.2021.100877
39. Zheng H, Cao X (2021) Impact of high-speed railway construction on spatial relationships in the Guanzhong Plain urban agglomeration. Reg Sustain 2(1):47–59
40. Zhou Y, Chen Y, Xie B, Pei T, Yi X (2019) Analysis on the connection and influence range of Guanzhong plain urban agglomeration. AR Res Dev 38(3):54–59. https://doi.org/10.3969/j.issn.1003-2363.2019.03.011
41. Zhu S (2018) Features of administrative division of Xi'an centered Guanzhong area in the history and inspiration for today. <历史时期以西安为中心的关中地区政区设置之特点及其当代启示>. From the Institute of Northwest Historical Environment and Economic and Social Development, Shaanxi Normal University, Available from: http://skxb.chd.edu.cn/oa/DArticle.aspx?id=000110&type=view
42. Zhu Y, Tong Q, Zeng X, Yan X, Li Y, Huang G (2019) Optimal design of a distributed energy system using the functional interval model that allows reduced carbon emissions in Guanzhong, a rural area of China. Sustainability 11(7):1930

Chapter 11
Central Plains Megalopolis: A Hub for Green Transformation and Development Beyond Just the Beijing-Guangzhou Channel

11.1 A Brief Introduction to the Central Plains Megalopolis

The Central Plains Megalopolis (CPM) is a large urban agglomeration that includes Zhengzhou, Kaifeng, Luoyang, Nanyang, Anyang, Shangqiu, Xinxiang, Pingdingshan, Xuchang, Jiaozuo, Zhoukou, Xin Yang, Zhumadian, Hebi, Puyang, Luohe, Sanmenxia, and Jiyuan in Henan Province. It also includes three cities of Changzhi, Jincheng, and Yuncheng in Shanxi Province and two cities of Xingtai and Handan in Hebei Province. Two cities of Liaocheng and Heze in Shandong and five cities of Huaibei, Bengbu, Suzhou, Fuyang, and Bozhou in Anhui Province are also included in the CPM city cluster. In total, this large city cluster includes 30 prefecture-level cities in five provinces. The total coverage area of the CPM is approximately 287,000 square kilometers stretched between Henan, Shanxi, Hebei, Shandong, and Anhui provinces [10]. In 2017, it had an estimated population of 163,531,700 and a total GDP of 6,777.812 billion Yuan [11].

The CPM urban agglomeration or city cluster is located in the intersection area of the land bridge passage and the Beijing-Guangzhou passage [21]. This nationwide passage is known as the strategic urbanisation pattern of "two horizontal and three vertical" [8, 51, 64, 81]. Across these five central provinces, high-speed railway networks and modern integrated transport hubs are accelerating [5, 23, 68]. The three-dimensional integrated transport network is constantly being improved to promote communication and transportation links [83] in a cross-provincial and nationwide attempt [14, 34, 72]. Industrial clusters such as equipment manufacturing and intelligent terminals have obvious regional advantages [1, 17, 27, 30]. Also, the logistics and tourism industries have a certain international influence, which plays a major part in the overall regional development plans.

The CPM urban cluster is the largest, most integrated, and most densely populated urban agglomeration within a radius of 500 km between the Changjiang River Delta, the Zhujiang Delta, and the Beijing-Tianjin-Hebei city cluster [26]. It is the hub and core area for the central region to undertake the industrial transfer of developed countries and the eastern region of China [43]. It is also known as the resource

A. Cheshmehzangi and T. Tang, *China's City Cluster Development in the Race to Carbon Neutrality*, Urban Sustainability, https://doi.org/10.1007/978-981-19-7673-5_11

export of the western region. The overarching plan for the region is to become the core growth pole to participate in domestic and international competition, promote the rise of the central region, and radiate the development of westen and central regions [60]. The main strategies are to promote the development of the CPM city cluster to a great strategic significance for accelerating the rise of the central region [15, 44], promoting the construction of new urbanisation [5, 6], and expanding the new space of economic development in China.

11.2 A Historical Analysis of the Central Plains Megalopolis

The 'Central Plains' region, also known as Huaxia, Middle Earth, and Zhongzhou, refers to the middle and lower reaches of the Yellow River from Luoyang to Kaifeng. In a narrow sense, it refers to today's Henan Province. When it is discussed internationally, the Central Plains generally refer to China [3, 20]. The Central Plains region has the largest number of dynasties, the longest history, and the most significant number of ancient capitals in China [2, 12, 19, 55]. More than 300 emperors built or moved their capitals here in more than 20 dynasties [4, 73]. Throughout history [45], the Central Plains region has always been China's political, economic, cultural, and transportation center.

Since ancient times, it has been said that "whoever wins the central plains wins the country" and "only by chasing the deer can the country be divided" [42]. Historical records or archaeological evidence show that the Central Plains region occupied four of the eight ancient capitals of major regimes for a long time [25, 84, 85]. It is the ancient capital of thirteen dynasties Luoyang, eight dynasties Kaifeng, seven dynasties Anyang, Zhengzhou, and Shang Dynasty, in addition to Shangqiu, Nanyang, Puyang, Xuchang, Dengfeng, Xiayi, Yanshi, Yucheng, Huaiyang, and Xinzheng.

Since the reform and opening up in 1978, China's regional development has gone through the process of coastal opening, the development of the west, the revitalisation of the northeast and other industrial bases, and the rise of central China [7, 37, 40, 80]. The rise of the central region is referred to the national level itself, allowing the country's regional development to form the basic pattern of four plates [63, 69]. As the largest province in the country and the central region, Henan has attached great importance to the rise of the Central Plains for its own development [33, 58, 62]. In 2003, Henan province put forward the economic development plan for "the rise of the central plains" [29, 87]. In this major development plan, the provincial government put forward a constructive approach for six key areas of 'advanced agriculture', 'advanced industry', 'prosperous culture', 'a beautiful environment', 'social harmony', and 'people's prosperity. These were identified as the general goal of realising the rise of the central plains. In 2016, the National Development and Reform Commission (NDRC) officially approved the "Central Plains Urban

Agglomeration Planning", which is now an ongoing national-level strategy for the regional development.

11.2.1 An Overview of Population Migration of the Central Plains Megalopolis

Shanxi is a major energy-producing province. The population of its cities is also relatively low compared to neighbouring regions. The permanent population of the province was 34,915,600 in 2020, compared with 35,712,100 in 2010 [88]. This shows a decrease of 796,500 people, or 2.23%, which indicates a population outflow at the regional level. As per governmental reports, nine of the 11 municipalities experienced a decline in their population over the same decade. In addition to Shanxi, Anhui, Hubei, Hunan, and other central provinces have all experienced similar large population decline [88]. For example, 10 out of 16 cities in Anhui experienced a decline in their population; eight in Hunan and Hubei (excluding county-level cities and forest areas directly under the provincial jurisdiction), and six in Henan and Jiangxi.

On the one hand, the central region is almost central to the Yangtze River Delta, Pearl Delta, and Beijing-Tianjin-Hebei, which are the three economic circles in China. In recent years, with the rapid development of the high-speed rail network, the hinterland of the three economic circles continues to extend, and the personnel exchanges between the central regions with abundant labor resources are more convenient [88]. Taking Henan as an example, compared with 10 years ago, the proportion of the outflow from Henan to Jiangsu, Zhejiang, and Shanghai increased by 5.1%, 2.3%, and 1.4%, respectively. Nearly 40% of Henan's migrant population has gone to Jiangsu, Zhejiang, and Shanghai, which is almost 10% more than the decade before it [88]. In addition to the flow to the southeast coastal developed areas, the province's floating population increased significantly in many central and western provinces. This mainly occurred from ordinary cities in the province to the provincial capital and central cities. This population mobility was more significant, especially in Henan, Sichuan, Anhui, Hubei, Yunnan, and other central and western provinces [49, 76]. In recent years, these regions have accelerated the development of the strong province and helped other cities in the province to speed up the flow of population.

The CPM city cluster is located in the middle and eastern parts of China, which is the largest and most densely populated urban agglomeration in China [16]. According to the fifth census data and the sixth census data for ten years, the CPM city cluster population ageing coefficient has risen from 7.09% to 8.48%. Thus, it has entered the aging society. Due to the large economic development gap of each basic regional unit and the different population bases, the population structure difference is noticeable [88]. To explore the spatio-temporal evolution of ageing population and its influencing factors, the CPM city cluster can provide reference to formulating relevant population policies, guiding rational population development, delaying ageing, and solving a series of problems caused by ageing population [88]. Also, in order to promote the coordinated development of the population and economy of the CPM city cluster, we note the high level of significance of the region.

The spatial distribution of ageing population type changes significantly, mainly from non-ageing and shallow ageing to deep ageing society. Meanwhile, from 2000 to 2020, the spatial distribution of population aging presents a trend of "southeast higher than northwest lower". By 2030, the spatial distribution of ageing population is expected to be higher in the northwest than in the southeast. Third, from the correlation pattern, the population ageing in the CPM city cluster has positive correlation characteristics from 2000 to 2030. The association type area is mainly H–H and L–L areas, which is a cluster distribution. From the center of gravity migration, the moving direction and distance of the total population center and the elderly population center of the CPM city cluster changed from 2000 to 2030. The total and elderly population centers are almost located in the southeast of the geometric center. But the change direction of the two is not consistent. Thus, the center of gravity of the total population continues to move northwest, while the elderly population center first moves southwest and then moves northwest [88]. From the pattern of concentration and sparse, the spatial distribution of population aging in the CPM city cluster shows that the north is higher than the south, the west is lower, and the east is higher.

11.3 An Overview of the Industrial Structure in the Central Plains Megalopolis

In the CPM city cluster, transportation location (e.g., nodes, networks, connections, etc.) is important. Located in the hinterland of China, i.e., connecting east and west and connecting south and north, the CPM city cluster is the intersection area of the land bridge channel and the Beijing-Guangzhou channel. As mentioned earlier, it is in the strategic pattern of "two horizontal and three vertical" urbanisation [22]. At the end of 2011, the mileage of railways and highways in the region reached 6,965 km and 8,323 km, respectively, accounting for 7 and 9.8% of the national total [45, 88]. Seven civil airports are in operation, making it an important hub in the national comprehensive transportation network.

The second aspect of that grain superiority is prominent in the region. Agricultural production conditions are superior, which suggests the range of important (main) production areas of agricultural products [70]. Grain output exceeded 100 million tons, accounting for more than 18% of the country's total, of which 54 million tons of wheat were produced, close to 50% of the total [88]. Cotton, oil, livestock, and poultry production accounted for 18.4, 20.5, and 14.8% of the country [45]. Thus, the characteristics of agricultural and forestry products occupy an important position in the country. They are also the backbone and structure of primary industries in the region.

The government reports indicate that the industrial foundation is suitable at the regional level. The CPM city cluster is rich in mineral resources, coal, aluminum, molybdenum, gold, trona (a monoclinic mineral), and other large reserves [46]. Also, it is an important source of energy and raw materials base [31, 81]. The region has a complete range of industries, with prominent advantages in the equipment, nonferrous metals, and food industries. The electronic information, automobile, light industry, and other industries have grown rapidly in scale, forming a relatively complete industrial system. In addition, the market potential is huge [47]. In recent years, the urbanisation rate reached 40.6 percent [88], while the country is in the stage of accelerating industrialisation and urbanisation. The rate is almost 20 percent lower than the national level. There is broad space for investment and consumption demand, and market advantages are increasingly apparent. The population is large, and the quality of the labour force continues to improve. The CPM city cluster is one of the country's most abundant labour resources It also supports neighbouring regions, particularly near Beijing and Shanghai [57]. The open economy has developed rapidly, and an all-directional opening pattern has gradually taken shape.

Lastly, the CPM city cluster has a profound cultural heritage [41, 75]. It is an important cradle of the Chinese nation and Chinese civilization. With a long history and a large amount of valuable historical and cultural heritage, it has formed a culture of inclusiveness, hardness and softness, innovation, and endless growth. The soft power of culture has been continuously enhanced at the regional and local levels. Nonetheless, the development of the CPM city cluster faces some outstanding contradictions and problems. China's economy continues to advance from east to west, but the pattern of east and west has not changed in general. In this regard, the phenomenon of regional economic trend differentiation and the increasing difference between north and south has become more prominent. The internationalisation degree of the central city is low, and the international gateway hub function and the high-end factor service function are insufficient. The number of medium to large cities with a population of more than one million in urban areas is insufficient in this region. Also, the level of modernisation of small and medium-sized cities is not high. Thus, we can see the development of science, technology, and education is uneven across the region. For instance, the overall capacity for independent innovation is not strong [86], and there is a shortage of high-quality and high-level personnel. Resource and environmental constraints have intensified, with smog and water shortage in some cities becoming more prominent in recent years. Finally, the system and mechanism of inter-city integration and coordinated development need to be further improved.

11.4 The CPM City Cluster's Sustainability Agenda: Plans and Perspectives

In regards to the sustainable development plans, the CPM city cluster adheres to the green, low carbon, sustainable development concept and puts the ecological environment construction in a prominent position [9]. The government plans to establish and improve cross-regional mechanisms for building an ecological civilization, resolutely win the battle against air, water, and soil pollution, and create a livable urban cluster with green mountains, clean water, and blue sky, where people and nature are in harmony [48, 65]. There are three significant directions at the regional level, including (1) building green cities, (2) developing an urban circular economy, and (3) promoting green and low-carbon urban development.

Building green cities

The overall plan is to improve the efficiency of urban resource utilisation. The cross-provincial governments plan to cooperate in the R&D and production of common key technologies and complete sets of equipment for energy conservation and emission reduction [28, 61, 79]. They jointly organise the implementation of key energy-saving technological transformation projects and achieve clean and efficient utilisation of coal and other energy sources in the region [52, 61]. They also implement dual controls on the total amount and intensity of water consumption, promote the system of leading water efficiency, popularise high-efficiency water-saving technical products and water-saving appliances, improve the availability of water measuring instruments, and build water-saving cities [36]. The strategies are to strengthen planning and demonstration of the distribution of water resources for major projects, implement the principle of using water to determine cities and production, and clarify targets for controlling the total amount of water used and for limiting pollution intake in water function zoning [35]. The governments at multiple levels aim to carry out afforestation, urban renewal, and projects to use waste heat for heating households. The plan is to tighten controls on land use and promote the redevelopment of low-efficiency land in cities and towns [67]. In doing so, future development focuses on the reclamation and utilisation of abandoned industrial and mining lands and encourage the development and utilisation of the urban underground spaces.

Develop an urban circular economy

The government aims to implement the circular development guiding plan, focusing on building circular industrial chains in the fields of non-ferrous metals, coal, non-metallic minerals, and renewable resources [31, 54, 74]. These can be done by actively creating national circular economy demonstration zones and promoting the construction of urban mineral demonstration bases [16, 50]. There is support for the construction of national circular economy demonstration cities (or counties) such as Hebi, Luoyang, and Xinxiang. The involved governments vigorously develop the remanufacturing industry, promote the construction of the recycling system of renewable resources, and gradually implement the classification and recycling treatment of urban garbage. Their primary strategies are to speed up the formation of a trans-regional recycling network, strengthen the recycling of low-value waste, and create a number of resource recycling bases. In doing so, they propose to promote the recycling of sewage treatment and bring recycled water into the unified allocation of regional water resources.

Promote green and low-carbon urban development

The government plans to strengthen total carbon emissions control. It also aims to effectively control carbon emissions in critical industries such as electricity, steel, building materials, and chemicals. The primary strategy is strengthening carbon sequestration capacity building and improving the green space, wetlands, and other carbon fixation levels. In doing so, the plan is to raise standards for building energy efficiency, promote green building and building materials, and promote energy-saving, livable and green renovation of old urban districts. These will strengthen the management of energy conservation in public buildings, improve energy use efficiency, and promote solar energy, shallow geothermal energy, and other renewable energy sources in line with local conditions. One of the key directions is to vigorously develop prefabricated buildings. This approach is proposed to actively respond to and actively adapt to climate change impacts, advocate a low-carbon lifestyle, and encourage green travel and consumption. In doing so, the government supports the construction of low-carbon cities, communities, industrial parks, and towns.

11.4.1 A Major Shift to Improve the Urban Infrastructure System

The first major shift is to strengthen the construction of municipal public facilities. In this regard, the government plans to speed up the urban road network system's improvement, increase the road network's density, and enhance accessibility. Central cities will be supported to advance the construction of public parking facilities and electric vehicle charging facilities in an orderly manner, focusing on residential areas, large-scale comprehensive transport hubs, and outlying urban rail transit stations [1]. The involved stakeholders are instructed to accelerate the transformation and upgrading of urban infrastructure and improve the quality of construction, operation standards, and management [59]. The associated aspects are to speed up the construction and renovation of urban water supply facilities, increase the penetration of public water supply, reduce the leakage of water supply networks, and support the extension of urban water-supply services to rural areas where conditions permit.

The second major shift is to speed up the construction of urban underground pipe corridors. The government plans to comprehensively promote the construction of comprehensive underground pipe corridors in large and medium-sized cities. It also aims to support cities that have the conditions to take the lead in building a comprehensive urban underground pipe corridor system that combines trunk and branch branches with cable pipe corridors as a supplement. Small cities are encouraged to advance in an orderly manner the renovation of overhead cables into corridors of existing urban power grids and communication networks. Basically, they eliminate the problem of spider-web overhead cables in primary roads and streets. The central government plans to speed up the construction of Zhengzhou as a pilot city for underground pipe corridors.

The third major shift is to give priority to the development of urban public transport. This plan is strategised through pilot cases, speeding up the construction of urban rail transit in Zhengzhou and Luoyang, and planning and studying the urban rail transportation in Handan [53, 64]. The government encourages cities with a population of 1.5 million or above to plan and build light rail systems, develop modern streetcars and other large-capacity public transportation, improve slow-moving transportation systems such as walking and cycling, and increase the proportion of trips made by public transportation and green transportation in central cities.

The fourth major shift is related to building sponge cities in the region. The government actively promotes the low-impact development (LID) model in the US and supports large and medium-sized cities in integrating the development of new urban areas with the renovation of shantytowns in old urban areas and the organic renewal of old residential areas [18, 24, 32, 77]. The construction and renovation of sponge cities should be promoted systematically, and small cities should give priority to the construction of sponge parks, green spaces, squares, and residential areas in new urban areas. One of the main pilot cases is to accelerate the construction of the Hebi National Sponge City pilot project. In such experimental and (later on) scaling up processes, the government plans to strengthen the construction of urban drainage and waterlogging prevention facilities. They propose to speed up the completion of urban drainage and waterlogging facilities short board, build a source of emission reduction system, drainage canal system, waterlogging relief, exceeding the emergency system as the core of the urban drainage system. In doing so, the plan is to fully meet the national waterlogging control standards.

11.5 Pathways to Carbon Neutrality in the Central Plains Megalopolis

One of the main plans aligned with the carbon neutrality plan is to have a carbon balance zoning of the CPM economic region at the county level [82]. Recent work focuses more on spatio-temporal evolution in the region [13], where the focus should be more on carbon source, carbon budget, and carbon sink. This, regional ecological efficiency is recognised as an influencing factor from the carbon neutrality perspective [56]. Likewise, we see the spatial–temporal approaches are important in developing net carbon sink efficiency [78], which are likely to be the focus of the regional development in CPM by 2030. The review of the agricultural and cropping systems is also part of the agenda for this decade, if not any longer. Once again, spatio-temporal innovations should be looked at from the sound spatial structure plans to ensure carbon emission reductions are aligned with the industrial structure and longer-term strategies of low carbon development, double carbon goal [66], etc.

One of the potential areas to focus on is the construction of forest cities aligned with low carbon development [39], which could help balance the carbon emissions from mixed industries. In this regard, a dual perspective of industry and space [38] is considered to be critical to spatial transformations, the industrial restructuring process, and meeting the carbon peak targets of 2030. Unlike most other urban agglomerations, there must be a lot of emphasis on the CPM's spatial spillover effects of environmental pollution at the regional level [71]. Thus, we note the starting point to reach the carbon peak and carbon neutral targets for the CPM city cluster is spatial strategies that should be correlated with industrial development, agricultural production, economic growth, and transportation networks. To make these into a reality, the CPM city cluster ought to consider large-scale spatial restructuring and planning interventions.

References

1. Bao X (2018) Urban rail transit present situation and future development trends in China: overall analysis based on national policies and strategic plans in 2016–2020. Urban Rail Transit 4(1):1–12
2. Bunker EC (1993) Gold in the ancient Chinese world: a cultural puzzle. Artibus Asiae 53(1/2):27–50
3. Cao B (2022) Miaodigounization and Erlitounization: the formation and evolution of the Hua-Xia ethnic group and Hua-Xia tradition from the perspective of archaeology. Int J Anthropol Ethnol 6(1):1–25
4. Chen Y, Lukash S, Qingran G (2019) Econometric analysis on the relationship between agri-tourism development and economic growth-the case of central plains economic zone of China. *Економіка* 33, No., 2, UDC 338.24. Available form: http://econommeneg.btsau.edu.ua/sites/default/files/visnyky/economika/economika_upravl_apk_2_2019.pdf#page=33
5. Cheshmehzangi A (2014) The urban and urbanism: China's new urbanization and approaches towards comprehensive development. Int J Interdiscip Environ Stud 8(3–4):1–12
6. Cheshmehzangi A (2016) China's New-Type Urbanisation Plan (NUP) and the foreseeing challenges for decarbonization of cities: a review. Energy Procedia 104:146–152
7. Cheshmehzangi A (2020) Towards urban energy management: a brief review of China's recent transition in urban energy demand. Low Carbon Economy 11(01):1
8. Cheshmehzangi A, Dawodu A, Sharifi A (2021) Sustainable urbanism in China. Routledge, New York
9. Cheshmehzangi A, Li H, Yang X (2018) Low carbon and climate resilient urban development in China. ADB Report, developed for the National Development Reform Committee (NDRC), Beijing, China
10. China Daily (2017) Megalopolis strategy spurs mid-west China latecomers. Available from: http://europe.chinadaily.com.cn/business/2013-12/06/content_17158442_2.htm
11. China Daily (2021) A new stage of urbanization: a look at China Integrated City Index. Available from: https://www.chinadaily.com.cn/a/202104/17/WS607a4d45a31024ad0bab631e.html

12. Dai LL, Li ZP, Zhao CQ, Yuan J, Hou LL, Wang CS, Fuller BT, Hu YW (2016) An isotopic perspective on animal husbandry at the Xinzhai site during the initial stage of the legendary Xia Dynasty (2070–1600 BC). Int J Osteoarchaeol 26(5):885–896
13. Dong Y, Jin G, Deng X, Wu F (2021) Multidimensional measurement of poverty and its spatio-temporal dynamics in China from the perspective of development geography. J Geog Sci 31(1):130–148
14. Du Y (2021) Transformation and revival: research on urban development strategic planning for HanDan under the dual dilemma of resource-based industry transition and regional spatial marginalization. Master Disseration at Delft University of Technology, the Netherlands. Available from: https://repository.tudelft.nl/islandora/object/uuid:3f183391-5bfb-4155-9733-89a7a77c5897
15. Fang C, Yu D (2020) China's urban agglomerations. Springer, Cham, pp 1–32
16. Fu X, Li L, Lei Y, Wu S, Yan D, Luo X, Luo H (2020) The economic loss of health effect damages from PM2. 5 pollution in the Central Plains Urban Agglomeration. Environ Sci Pollut Res 27(20):25434–25449
17. Guo T, Wang Y (2019) Big data application issues in the agricultural modernization of China. Ekoloji 28(107):3677–3688
18. Guo X, Zhang J, Zhou B, Liu W, Pei J, Guan Y (2019) Sponge roads: the permeable asphalt pavement structures based on rainfall characteristics in central plains urban agglomeration of China. Water Sci Technol 80(9):1740–1750
19. He N (2018) Taosi: an archaeological example of urbanization as a political center in prehistoric China. Archaeol Res Asia 14:20–32
20. Huang W, Xi M, Lu S, Taghizadeh-Hesary F (2021) Rise and fall of the grand canal in the ancient Kaifeng city of China: role of the grand canal and water supply in urban and regional development. Water 13(14):1932
21. Jiao J, Zhang F, Liu J (2020) A spatiotemporal analysis of the robustness of high-speed rail network in China. Transp Res Part D: Transp Environ 89:102584
22. Jin F, Wang C, Li X, Wang JE (2010) China's regional transport dominance: density, proximity, and accessibility. J Geog Sci 20(2):295–309
23. Jin F, Ding J, Wang JE, Liu D, Wang C (2012) Transportation development transition in China. Chin Geogra Sci 22(3):319–333
24. Jin M, Lancia M, Tian Y, Viaroli S, Andrews C, Liu J, Zheng C (2021) Hydrogeological criteria to improve the sponge city strategy of China. Front Environ Sci 238
25. Jing Y, Campbell R, Castellano L, Xianglong C (2020) Subsistence and persistence: agriculture in the central plains of China through the Neolithic to Bronze age transition. Antiquity 94(376):900–915
26. Kang P, Song G, Chen D, Duan H, Zhong R (2020) Characterizing the generation and spatial patterns of carbon emissions from urban express delivery service in China. Environ Impact Assess Rev 80:106336
27. Kenderdine T (2017) China's industrial policy, strategic emerging industries and space law. Asia Pac Policy Stud 4(2):325–342
28. Kong F, Zhang H, Meng X, Li S, Liu J (2022) Can the policy of national urban agglomeration improve economic and environmental gains? evidence from Quasi-natural experiments with 280 cities in China. Int J Environ Res Public Health 19(13):7596
29. Lai H (2007) Developing central China: a new regional programme. China: Int J 5(1):109–128
30. Li L, Ma S, Zheng Y, Xiao X (2022) Integrated regional development: Comparison of urban agglomeration policies in China. Land Use Policy 114:105939
31. Li L, Shan Y, Lei Y, Wu S, Yu X, Lin X, Chen Y (2019) Decoupling of economic growth and emissions in China's cities: a case study of the central plains urban agglomeration. Appl Energy 244:36–45
32. Li R, Zhang C (2020) Selection and application of garden plants in the construction of Sponge City in Northwest China. J Coast Res 103(SI):1139–1143
33. Li X (2018) Research on the predicaments and countermeasures for the modernization of central plains' traditional culture. In: At the 4th international conference on economics, management

and humanities science (ECOMHS 2018). Available from: https://webofproceedings.org/proceedings_series/ECOM/ECOMHS%202018/ECOMHS97013.pdf
34. Li X, Liu X (2018) Green development model of China's small and medium-sized cities. Springer, Singapore
35. Li Y, Yin S, Yu S, Bai L, Wang X, Lu X, Ma S (2021) Characteristics of ozone pollution and the sensitivity to precursors during early summer in central plain, China. J Environ Sci 99:354–368
36. Li Y, Yuan X, Tang Y, Wang Q, Ma Q, Mu R, Fu J, Hong J, Kellett J, Zuo J (2020) Integrated assessment of the environmental and economic effects of "coal-to-gas conversion" project in rural areas of northern China. Environ Sci Pollut Res 27(13):14503–14514
37. Liang L, Chen M, Luo X, Xian Y (2021) Changes pattern in the population and economic gravity centers since the reform and opening up in China: the widening gaps between the South and North. J Clean Prod 310:127379
38. Liao B, Li L (2022) How can green building development promote carbon emission reduction efficiency of the construction industry? Based on the dual perspective of industry and space. Environ Sci Pollut Res 29(7):9852–9866
39. Liao L, Zhao C, Li X, Qin J (2021) Towards low carbon development: The role of forest city constructions in China. Ecol Ind 131:108199
40. Lin GC (2002) The growth and structural change of Chinese cities: a contextual and geographic analysis. Cities 19(5):299–316
41. Liu X (2018) International communication of intangible cultural heritage in central plains: a case study of Chinese Wushu. Int J Soc Sci Humanities 2(3):196–204
42. Liu Y (2021) Spread of artistic spirit of chinese historical drama songs. Tob Regul Sci 7(5):3652–3663
43. Liu YJ (2014) Urban spatial strategic research and its practice in China-a case study of the central plain cities. In: Advanced materials research, vol 838. Trans Tech Publications Ltd., pp 3050–3053
44. Long Y, Huixuan H (2013) The adjustment of China's regional development strategy and its impact on intergovernmental relations. In: Mierzejewski D (ed) Dimensions of Development. East Asia in the process of changes. Wydawnictwo Uniwersytetu Łódzkiego, Łódź 2013; Wydawnictwo Uniwersytetu Łódzkiego
45. Lukash SM, Yang C (2019) Econometric analysis on the relationship between agritourism development and economic growth-the case of central plains economic zone of China. Available from: http://repo.snau.edu.ua/handle/123456789/8387
46. Ministry of Mineral Resources (2020) China mineral resources. Available from: https://www.mnr.gov.cn/sj/sjfw/kc_19263/zgkczybg/202010/P020201022612392451059.pdf
47. Naughton BJ (2018) The Chinese economy: adaptation and growth. MIT Press, Cambridge
48. Pan J (2016) China's environmental governing and ecological civilization. Springer, Berlin Heidelberg
49. Pan J, Lai J (2019) Spatial pattern of population mobility among cities in China: case study of the national day plus mid-autumn festival based on tencent migration data. Cities 94:55–69
50. Qi J, Zhao J, Li W, Peng X, Wu B, Wang H (2016) Development of circular economy in China. Springer, Singapore, p 273
51. Ramondetti L (2022) The enriched field: urbanising the central plains of China. Birkhäuser, Basel
52. Shi J, Huang W, Han H, Xu C (2021) Pollution control of wastewater from the coal chemical industry in China: environmental management policy and technical standards. Renew Sustain Energy Rev 143:110883
53. Song X, Hao Y (2021) Research on the vehicle emission characteristics and its prevention and control strategy in the central plains urban agglomeration, China. Sustainability 13(3):1119
54. Su B, Heshmati A, Geng Y, Yu X (2013) A review of the circular economy in China: moving from rhetoric to implementation. J Clean Prod 42:215–227
55. Su Y, Liu L, Fang XQ, Ma YN (2016) The relationship between climate change and wars waged between nomadic and farming groups from the Western Han Dynasty to the Tang Dynasty period. Clim Past 12(1):137–150

56. Tan J, Wang R (2021) Research on evaluation and influencing factors of regional ecological efficiency from the perspective of carbon neutrality. J Environ Manage 294:113030
57. Tian Y, Jiang G, Zhou D, Ding K, Su S, Zhou T, Chen D (2019) Regional industrial transfer in the Jingjinji urban agglomeration, China: an analysis based on a new "transferring area-undertaking area-dynamic process" model. J Clean Prod 235:751–766
58. Wan H (2016) Study on tourism of Henan Wushu culture and its sustainable development under construction of central plains economic zone. In: International conference on education, management and computing technology (ICEMCT-16). Atlantis Press, pp 428–432
59. Wang DG, Niu Y, Qian J (2018) Evolution and optimization of China's urban tourism spatial structure: a high speed rail perspective. Tour Manage 64:218–232
60. Wang J, Liu S (2018) SWOT Analysis of the development planning of central plains urban agglomeration under the background of new era. In: The 4th international conference on innovative development of E-commerce and logistics (ICIDEL 2018). Available from: http://clausiuspress.com/conferences/LNEMSS/ICIDEL%202018/EL10563.pdf
61. Wang L, Sun H, Hu X, Leng Z, Zha Z, Zhang Z, Yang J (2021) Measurement of China's provincial consumption-based PM2. 5 emissions and its influencing factors in the perspective of spatial heterogeneity. J Clean Prod 317:128367
62. Wang X (2016) Empirical analysis of relation between export trade and economic growth in Henan Province. In: 2016 international conference on education, sports, arts and management engineering. Atlantis Press, pp 264–269
63. Wang X, Fan G (2004) Analysis on the regional disparity in China and the influential factors. Econ Res J 1(4):33–44
64. Wang X, Zhang W (2019) Efficiency and spatial equity impacts of high-speed rail on the central plains economic region of China. Sustainability 11(9):2583
65. Wang Y (2019) The detection method of rural environmental pollutants based on ecological civilization. Ekoloji 28(108):1481–1485
66. Wang Y, Niu Y, Li M, Yu Q, Chen W (2022) Spatial structure and carbon emission of urban agglomerations: spatiotemporal characteristics and driving forces. Sustain Cities Soc 78:103600
67. Wang Y, Zhang D, Ji Q, Shi X (2020) Regional renewable energy development in China: a multidimensional assessment. Renew Sustain Energy Rev 124:109797
68. Wei H, Nian M, Li L (2020) China's strategies and policies for regional development during the period of the 14th five-year plan. Chin J Urban Environ Stud 8(02):2050008
69. Wu F (2015) Planning for growth: urban and regional planning in China. Routledge, New York
70. Wu Z, Liu M, Davis J (2005) Land consolidation and productivity in Chinese household crop production. China Econ Rev 16(1):28–49
71. Xiong L, De Jong M, Wang F, Cheng B, Yu C (2018) Spatial spillover effects of environmental pollution in China's central plains urban agglomeration. Sustainability 10(4):994
72. Xiao L (2016) New patterns of global governance and China's strategy of free trade zone. In: National test. Springer, Singapore, pp 353–419
73. Xu H (2022) Genius of the central plains. In: The Earliest China. Springer, Singapore, pp 61–74
74. Xue B, Chen XP, Geng Y, Guo XJ, Lu CP, Zhang ZL, Lu CY (2010) Survey of officials' awareness on circular economy development in China: based on municipal and county level. Resour Conserv Recycl 54(12):1296–1302
75. Yan L, Yang R, Lu P, Teng F, Wang X, Zhang L, Chen P, Li X, Guo L, Zhao D (2021) The spatiotemporal evolution of ancient cities from the late Yangshao to Xia and Shang Dynasties in the Central Plains, China. Herit Sci 9(1):1–16
76. Yang Z, Gao W, Zhao X, Hao C, Xie X (2020) Spatiotemporal patterns of population mobility and its determinants in Chinese cities based on travel big data. Sustainability 12(10):4012
77. Yuan Y, Zhang Q, Chen S, Li Y (2022) Evaluation of comprehensive benefits of sponge cities using meta-analysis in different geographical environments in China. Sci Total Environ 836:155755
78. Zhang A, Deng R (2022) Spatial-temporal evolution and influencing factors of net carbon sink efficiency in Chinese cities under the background of carbon neutrality. J Clean Prod 365:132547

79. Zhang J, Wang J, Yang X, Ren S, Ran Q, Hao Y (2021) Does local government competition aggravate haze pollution? a new perspective of factor market distortion. Socioecon Plann Sci 76:100959
80. Zhang P (2008) Revitalizing old industrial base of Northeast China: process, policy and challenge. Chin Geogra Sci 18(2):109–118
81. Zhao D, Lei Y, Zhang Y, Shi X, Liu X, Xu Y, Xue W (2022) Analysis of vehicular CO_2 emission in the central plains of China and its driving forces. Sci Total Environ 814:152758
82. Zhao RQ, Zhang S, Huang XJ, Qin YC, Liu Y, Ding ML, Jiao SX (2014) Spatial variation of carbon budget and carbon balance zoning of central plains economic region at county-level. Acta Geogr Sin 69(10):1425–1437
83. Zhou G, Zhao C, Wu M, Chen Y, Zhou F (2022) Spatial heterogeneity of coupling coordination development between logistics and economy in central plains of China. Discret Dyn Nat Soc
84. Zhou L (2016) From state to empire: human dietary change on the Central Plains of China from 770 BC to 220 AD. Archaeol Res Asia 20:1–8
85. Zhou L, Yang S, Han Z, Sun L, Garvie-Lok SJ (2019) Social stratification and human diet in the Eastern Zhou China: An isotopic view from the Central Plains. Archaeol Res Asia 20:100162
86. Zhu W, Yue Z, He N, Luan K, Ye L, Qian C (2022) Analysis of China's urban innovation connection network evolution: a case study of Henan Province. Sustainability 14(3):1089
87. Zhu Y, Shi Y, Jia M (2017) Research on high-end manufacturing and aviation economy development of Henan Province. In: MATEC web of conferences, vol 100. EDP Sciences, p 01009
88. ZYJJW (2021) China's Ten Years Of Population Migration! 149 cities with a declining population, where they went. <中国十年人口大迁徙!149市人口减少, 他们去了哪里>. Available from: https://www.zyjjw.cn/news/china/2021-08-10/692430.html

Chapter 12
Beibu Gulf City Cluster: A New Strategic Pivot to Foster New Growth Drivers

12.1 A Brief Introduction to the Beibu Gulf City Cluster

The Beibu Gulf City Cluster (BGCC) in comprised of many cities in the Guangxi Zhuang Autonomous Region, such as Nanning, Beihai, Qinzhou, Fangchenggang, Yulin, and Chongzuo. It also includes cities from the Hainan Province, including four cities of Zhanjiang, Maoming, Yangjiang, and Haikou, and four countries of Danzhou, Dongfang, Chengmai, Lingao, and Changjiang. It is commonly known as the Guangxi-Hainan city cluster. It covers a total land area of 116,600 km^2 [38]. In 2021, it had an estimated population of 43.95 million and a total GDP of 2,144.83 billion.

The overall positioning of the BGCC is to make full use of geographical advantages. It taps on the regional characteristics and builds a blue Gulf city group oriented to the ASEAN region [56], and serving the "three south" concept (i.e., southwest, south, south, and south). It offers a vast region of livable and suitable cities for the industry. Centering on the overall positioning, the BGCC aims to accelerate the realisation of breakthroughs in the following development positioning:

An important hub for the ASEAN international corridor

The government gives full play to the advantages of connectivity with ASEAN countries by sea and land, strengthening the construction of the China-Indo-China Peninsula land international corridor. It is also known as part of the Pan-Beibu Gulf Economic Cooperation [21]. Therefore, it is an urgent need to improve the major maritime transport corridor starting from the port cluster in the Beibu Gulf [37], build an international information corridor between China and the ASEAN region [25, 31], and create a hub for the major international corridor with convenient links with the ASEAN countries.

A new strategic pivot for the "three—south" opening up and development

The government gives full play to the role of the "three south" region as an essential gateway to the sea, facilitating rapid and multi-directional connectivity with urban agglomerations such as Chengdu-Chongqing, central Guizhou, central Yunnan, and the middle reaches of the Yangtze River. It also promotes the all-round deepening of exchanges and cooperation between inland provinces and the ASEAN countries [15, 53]. Thus, it is urgently needed to

A. Cheshmehzangi and T. Tang, *China's City Cluster Development in the Race to Carbon Neutrality*, Urban Sustainability, https://doi.org/10.1007/978-981-19-7673-5_12

strengthen coordinated development with the Guangdong-Hong Kong-Macao Greater Bay Area and the Pearl River and Xinjiang Economic Belt [9]. This development plan helps optimise regional cooperation mechanisms and accelerate the formation of a new pattern of coordinated, interactive, and complementary development in the eastern, central, and western regions.

Moreover, the 21st-Century Maritime Silk Road is an important gateway for the organic connection between the two routes. Relying on its unique geographical advantages, the government plans to build a new north-south land corridor linking the western part of China with the Indo-China Peninsula, strengthen the role of the Beibu Gulf port group in connecting domestic and international transportation, and promote the convergence, docking, convergence, and interaction of the 21st-Century Maritime Silk Road and the Economic Belt [2]. Thus, it is necessary to speed up efforts to build pilot bases for international production capacity cooperation [36], important service platforms, and links for people-to-people exchanges for countries along the Belt and Road Initiative (BRI) at the regional and sub-regional levels.

An important green industrial base in China

In accordance with the requirements of greening stock and increasing high-end development, the government focuses on providing green agricultural and seafood products [6, 27], high-value-added manufactured products, and eco-tourism products [44, 57]. It aims to eliminate and transfer a number of industries that do not meet the overall positioning of the city cluster [55], speed up the construction of a group of characteristic agricultural bases, demonstration zones for circular industries, and modern service industry clusters [41]. More importantly, it is planned to realise the green and intelligent development of the port industry [34] and build an industrial system that meets the environmental requirements of the Bay Area.

Land and sea coordinated development demonstration area

The government aims to make coordinated efforts to advance infrastructure construction, industrial and urban distribution, population distribution, natural shoreline utilisation, and marine environmental protection [19, 35]. They propose to coordinate the management of rivers flowing into the sea, the protection of inshore waters, the development of offshore waters, and the utilisation of offshore resources [63]. The plans are also to promote sound interaction between the land economy and the marine economy, build a service support base for the development and utilization of the South China Sea [16], and explore a new development model in which land and sea are coordinated, and people live in harmony.

12.2 An Overview of the Cluster's Significance Beyond a New Strategic Pivot

Superior resource factor endowment is an important factor in the regional development plan. The BGCC is located in the tropical and subtropical zones, sitting on the largest bay in southern China. The ecological environment quality is first-class in China, with rich resources in ports, coastlines, oil and gas, agriculture and forestry, and tourism [2, 17, 59]. The flat terrain provides the potential for land development while land utilisation is abundant. The environmental capacity is large, and the economic carrying capacity of the population is robust, making it an important strategic pivot at the national level.

The other factor is that development vitality is increasing at the regional level. In recent years, the economic growth of the BGCC has been consistently above the national average level [2]. The marine economy, leisure tourism, other characteristic industries [51], and port-based industrial clusters are gradually forming. Innovation and entrepreneurship are constantly emerging [42], human resources are relatively abundant, and the economy's overall strength is growing.

In recent years, we see the opening up and cooperation have been deepened. The opening of sea channels, starting with the Beibu Gulf port group, and the opening of land channels supported by border ports have accelerated [2]. The construction of such open platforms, such as the China-ASEAN Expo [26, 30], the Hainan International Tourism Island, the key development and opening up pilot zone, the border economic cooperation zone [45, 58], and the China-Malaysia "two countries and two gardens" [40, 54] advanced in an orderly manner. In sum, the fields of opening up and cooperation are expanding, and the open economy is taking shape.

At the regional level, there is evidence that urban development has a sound foundation. For instance, the City of Nanning has developed into a medium-sized city with over 3 million population. Haikou, Zhanjiang, and other cities have gradually played a leading role in urban development. Cities and towns in the south (ning), north (hai), qin (zhou), fang (chenggang), zhan (jiang), mao (ming), yang (jiang), hai (kou), mai (mai), and wen (chang) are relatively dense [2]. Other small and medium-sized cities and towns are growing faster. In this regard, we can point out that tropical and subtropical cities have obvious features, and infrastructure is improving across the region.

Lastly, a major advantage is the existing close social and cultural ties between different nodes of the region. The cities of the BGCC share the same culture, affinity, and similar folk customs, with close cultural and people-to-people exchanges and a strong sense of regional identity [3, 5, 23]. Cooperation among Guangdong, Guangxi, and Hainan provinces (or regions) in marine, tourism, and ecological management has been strengthened [61], and cooperation among neighboring regions have gradually advanced. Thus, the BGCC has turned into a major strategic pivot or zone.

12.2.1 Summary of Problems in the Beibu Gulf City Cluster

The overall level of economic development is not high in the BGCC. For instance, per capita GDP is below the national average [34]. In some regions where population density and economic density are lower than those of neighbouring countries, the development and transformation of traditional industries are facing serious difficulties [52]. Also, the advanced manufacturing and modern service industries are lagging behind [20], and the transition from old to new growth drivers [11] needs to be further accelerated.

The other problem is that the functions of central cities need to be strengthened. There are no large or megacities with strong radiation and leading role in the region. Nanning and other central cities have insufficient agglomeration and radiation effects

[32, 47]. They are not strong enough to attract high-end production factors. Innovation and entrepreneurship are not active enough [39], thus, cities' liveability level needs to be improved. Their role as hubs for opening up to the outside world remains to be enhanced. Therefore, it is apparent that the comprehensive competitiveness of cities still needs to improve.

At the regional level, we note cities are few and not well connected. There are only 22 cities and 538 towns in this urban agglomeration, with only 48 towns per 10,000 km^2. More importantly, the urban distribution is sparse. The heavy chemical industry dominates the port cities, and their economic relations are not close enough. In this regard, the ports have disorderly competition [2]. The inter-provincial cities lack efficient and convenient transportation channels and industrial structure convergence [45]. As a result, the division of labour is not enough, and the characteristics of isolated development are obvious.

Across the region, infrastructural development is a fundamental problem that requires improvement. Infrastructure investment is insufficient, and debts are high [7, 50, 60]. There are a few high-grade transport corridors, a few airport routes are set up, and communication channels with domestic and foreign countries are relatively poor. The intercity transport network is not sound [12, 15, 16], and connectivity and transport services need to be improved.

Lastly, we note that in recent years environmental constraints have been tightening in the Bay Area. The Beibu Gulf is a semi-enclosed bay with a weak current, which is not conducive to the diffusion and degradation of pollutants [22]. Pollution in inshore waters is rising, and the risk of marine environmental pollution is increasing [34]. In addition, the degradation of ecosystem services has not yet been fundamentally checked [32, 41], and the contradiction between development and protection is becoming increasingly prominent.

12.3 Summary of Strategic Opportunities in the Beibu Gulf City Cluster

Further progress has been made in building the Belt and Road Initiative (BRI), and the construction of the upgraded China-ASEAN Free Trade Area has progressed smoothly [8, 28]. In this regard, the BGCC gives space to its unique geographical advantages [2]. It benefits from the all-around expansion of opening up and promotes development by providing more space. The further implementation of the national strategy of new-type urbanisation and the development of the western region has provided a more decisive impetus for the BGCC to exert policy effects, make various cities bigger and stronger, and expand new space for new development [29, 39, 46]. The upgrading of the domestic consumption structure and the optimisation of the supply system provide a broader market for the Beibu Gulf region to play its ecological advantages and develop a green economy [2]. The Pearl River Delta and other developed areas have entered a new stage of industrial transformation and upgrading [4, 43, 48], which has provided more robust support for the Beibu Gulf region to play its latecomer advantage, undertake industrial transfer, and consolidate the industrial base.

A major regional strategic opportunity is to build green cities or promote green development. This strategy is based on the close integration of urban construction with the protection of landscape, forest, fields, and lake systems. The integration of coastal, landscape, pastoral, cultural, and other elements, urban production, living, and ecological space layout, help to create a harmonious relationship between humans and nature of the tropical and subtropical green city development [2]. Aligned with this strategy, the government promotes the construction of forest cities, strictly protects urban water systems, water bodies, and green centers, and weaves water networks into corridors and water into green spaces [41]. The government also currently carries out ecological restoration, urban repair, and transformation and promotes urban construction [59]. Therefore, it is necessary to accelerate the transformation and upgrading of urban industries, promote the circular transformation of industrial parks, and promote green and clean production [2]. Subsequently, the government aims to develop green energy, promote green buildings, prefabricated buildings, and green building materials, and build a green transportation system.

Another regional strategy is to invest in building sponge cities. This approach will strengthen the construction of water conservancy infrastructure and sponge cities, and improve the cities' ability to absorb, store, seep, and purify water. In this regard, the government pushes ahead with the construction of sponge cities and popularises the models of sponge buildings and communities, roads and squares, parks, and green spaces [2]. The overarching aim is to restore and enhance the function of urban sponge, local absorption and utilisation of Rain Water resources, and remediation of urban built-up areas of polluted water [2]. The secondary strategy is to actively promote the construction of an urban underground comprehensive pipe gallery, strengthen the pipeline into the gallery management, and improve the urban drainage capacity.

The third strategy is to build smart cities. Relying on the China-ASEAN Information Port, the government proposes accelerating the application of new generation information technologies such as big data, cloud computing, and the Internet of Things (IoT) in urban construction management [33]. They plan to promote informationization in urban planning, land use, underground pipelines and space management, landscaping, building management, and protection of historical and cultural blocks. This approach speeds up the construction of a time–space information cloud platform for smart cities [2]. Thus, the government vigorously develops smart municipal services, smart transportation, smart ports, smart energy, smart communities, and smart parks, and promotes the construction of smart cities such as Nanning, Haikou, Zhanjiang, Beihai, Qinzhou, Yulin, and Danzhou.

The fourth strategy is to build humanistic cities. The plan is to value and integrate the beautiful coastal sceneries, colourful ethnic customs, unique border advantages, rich history, and culture into the city construction [24]. Coastal cities should highlight coastal elements such as seawater, sand, sunshine, and harbors. Cities along the border should highlight customs and the ASEAN elements or characteristics. Also, cities inhabited by ethnic minorities should highlight ethnic elements. In this regard, the government proposes building many famous historical and cultural cities, towns, and neighbourhoods, and shaping the urban cultural brands.

The last strategy is to build an environmentally-friendly modern industrial system and adhere to the green and characteristic development direction. This strategy will turn resource advantage into an industrial advantage. It will also promote industrial development by opening up across the board, transforming and upgrading traditional industries, vigorously developing the marine economy, and fostering new growth drivers [2]. In this regard, regional development should promote industrial upgrading by gathering, and intensive clusters foster a number of industrial agglomeration areas with distinctive advantages. In doing so, the government should create a batch of new industrial demonstration bases [49] and comprehensively consolidate the industrial foundation of the urban agglomeration.

12.4 Sustainable Development Plan: A Multifaceted Approach in the Beibu Gulf City Cluster

A major sustainable development plan of the BGCC is to cultivate 'Green Industrial Clusters' to promote the green transformation of the Lingang Industry. The government plans to promote the transformation by relying on Qinzhou, Fangchenggang, Tieshan, Zhanjiang, Maoming, Yangpu, and other port areas and key industrial parks [22] and upgrading traditional industries. In doing so, it will build a green manufacturing system, realise the all-around development of green products, green factories, green parks, and green supply chains, and build an eco-industrial cluster in the Bay Area with substantial influence in China [2]. This development plan also promotes the integration of the refining and chemical industry, develops medium and high-end petrochemical products, improves the industrial chain, and builds a competitive petrochemical industry cluster. Green transformation of energy bases in Beihai, Qinzhou, Fangchenggang, Danzhou (Yangpu), Dongfang, and Yangjiang will be promoted in this plan. The secondary directions are to promote the clean transformation of 17 metallurgical industries and build Baosteel Zhanjiang base as the leading metallurgical deep processing industry cluster.

Another sustainable development plan is to foster emerging industrial clusters at multiple spatial levels. Relying on the advantages of land and sea resources, the government aims to jointly build several innovative parks, marine industrial parks, strategic emerging industrial bases, as well as press, publishing, radio, film, and television industry bases (parks). They also aim to foster emerging industrial clusters such as electronic information, high-end equipment, marine industry, new energy, biology, and geographic information [10]. With Nanning, Haikou high-tech industrial park, and Beihai industrial park as the core, the plan is to build an electronic information industry cluster. In doing so, it is necessary to promote the concentration of heavy machinery, shipbuilding, and petrochemical equipment manufacturing in port areas to build high-end equipment manufacturing bases. Thus, the proposals are to promote the construction of the Zhanjiang National Marine Hi-Tech Industrial Base and Beihai Marine Industry Science and Technology Park and build new energy bases in Beibei, Fangchenggang, Qinzhou, and Yangjiang [2]. Relying on the national high-tech biological industry base in Nanning and the national breeding and seed production base in southern China, the government aims to speed up the development of biological industries such as biological medicine and biological manufacturing.

The other regional sustainable development plan agenda is to build modern service industry clusters. Focusing on trade logistics, information services, modern finance, business and exhibition, marine services, and cultural services [28], Nanning, Haikou, and other cities will promote the agglomeration development of modern service industries. The government aims to enhance the logistics hub function of Nanning and build comprehensive trade logistics bases in Qinzhou, Fangchenggang, Zhanjiang, Yulin, and Maoming. The associated strategy is to enhance the cross-border financial services function of Nanning and Zhanjiang, promote regional financial reform and innovation, and study and carry out the pilot of offshore finance in Hainan. In this regard, the government proposes strengthening the brand effect of the China-ASEAN Expo and other exhibitions, upgrading the product exchange center of developing countries and the ASEAN region, and building an exhibition system with Nanning and Haikou as the centers and other cities as the nodes. Thus, the plan is to develop a big healthy industry, create a rich coastal development, and improve the health of landscape characteristics in the region.

The development of a special agricultural seafood processing cluster is a regional development plan that requires an array of sustainability measures and indicators. The plan will promote advanced and practical equipment, comprehensive utilisation, and quality and safety technologies. It also promotes the integrated development of raw material guarantees, food processing, and logistics marketing. Primary aspects of this process include the extension of the processing chain of agricultural and seafood products. It requires the involved stakeholders and governments to jointly build processing bases for sugar, grain and oil, beverage (mainly tea), meat and poultry, aquatic products, forest fruits, and other agricultural and forestry products. The project would initially build a green agriculture industry cluster. The government would foster and expand leading enterprises in agricultural industrialisation [2], create a whole chain of marine industries and modern agriculture, and promote the integrated development of primary, secondary, and tertiary industries in rural areas [28]. The ongoing construction of the Leizhou Peninsula tropical and subtropical modern agricultural demonstration area is part of this agenda.

As a supporting strategy, the plan is to build international tourism and leisure destinations in the BGCC jointly. The government aims to accelerate the development of leisure and vacation, eco-tourism, border tourism, ethnic customs, health and wellness, rural tourism, and red tourism (i.e., a subset of tourism in which people visit particular areas and sites with historical significance to communism in China). Thus, it is necessary to promote the transformation and upgrading of the tourism industry and improve the quality of international tourism services in urban agglomeration. The current focus is on building Hainan International Tourism Island, Beihai Yintan

National Tourism and Holiday Resort, Haihua Island Tourist Holiday, Yangjiang Hailing Island's Great Cape Bay Maritime Silk Road Tourist, and Zhanjiang's "five islands and one bay" coastal tourism demonstration zone [2]. Aligned with the regional development plans, the government promotes the construction of cross-border tourism cooperation zones such as Dongxing-Mangjie and Detian-Banyo. It builds a tourism development pattern of cooperation among large, medium, and small cities. In this process, the government explores the unique cultural resources of Luo Yue, Zhuang, Li Miao, Baiyue, Dan-Jia, and Hakka. This strategy is also aimed to enhance the existing transportation links. For instance, it is proposed to strengthen the transportation links of Nanning, Beihai, Chongzuo, Zhanjiang, and Haikou and build special tourism routes around the Bay [2]. These would lead to secondary development measures, such as integrating tourism resources along the coastline and the South China Sea island tourism resources and the development of a boutique cruise yacht line. The longer-term plan, perhaps to be continued differently after the COVID-19 pandemic era, is to promote visa-free and visa-on-arrival policies for specific countries and regions in key cities. Part of this strategy was successfully implemented before the pandemic in early 2019.

Based on its strategic location, one of the key development plans is to develop and strengthen the civil-military integration industry. In accordance with the requirements of military and civilian integration, land and sea coordination, resource sharing, and agglomeration development, this plan relies heavily on the advantages of the Nanning, Qinzhou, and Fangchenggang industry innovation platforms. This programmes aims to Build Beidou satellite navigation technology industrial park and promote the application of satellite navigation system core 19 industrialisation components [2]. The impacts are specifically on certain areas of significance in the region. For instance, the plan is to build an aerospace industry base in Haikou, Zhanjiang, and Maoming. The government aims to jointly build a civil-military integration public service platform, and promote the construction of several civil-civilian integration innovation incubation centers, public innovation spaces, and industrial demonstration parks. It is also proposed to promote the opening and sharing of military-civilian science and technology resources and the mutual transformation of technologies [2] that can be used for both military and civilian purposes.

From the economic perspective, one of the overarching plans is to create a sound ecosystem for innovation and entrepreneurship. At the regional development, this programme aims to integrate public service resources and explore the establishment of a three-in-one service model of government guidance, intermediary services, and social participation [2]. The government proposes to actively develop public services such as entrepreneurship guidance, information consulting, technical support, and trading of achievements. Cities with the conditions may explore and provide and standardise development financing guarantee services. Subsequently, the plan is also to build Nanning Zhongguancun Center for Entrepreneurship and Innovation, Zhanjiang Weishu Space for Innovation and Entrepreneurship, Haikou Renaissance City Internet Entrepreneurship Base, and Yulin Small and Micro Enterprise Innovation Base [2]. The programme encourages the development of new entrepreneurial spaces, such as maker spaces and venture coffee, improves investment and financing models

[2], and strengthens support for innovation and entrepreneurship. Aligned with this evolution in business and trade considerations, the aim is to cultivate and introduce innovative and entrepreneurial talents. Socially speaking, this is essential for the longer-term regional development plans. Thus, it is required to implement a more active talent policy and provide more convenient working and living conditions for all kinds of talents. The aim is to vigorously introduce international scientific and technological talent and a high-level innovation team and attract foreign talent and domestic talent to the BGCC's 23 areas of innovation and entrepreneurship. Currently, the government explores and builds an international cooperation talent training model and actively trains highly skilled personnel, professional managers, and middle managers. This programme aims to improve the incentive mechanism for talents, improve the two-way flow mechanism for scientific research talents, and fully stimulate the vitality of talents.

Lastly, as part of the regional development, the provincial and local governments emphasise guiding migrant workers to return home and start businesses. The plan is to implement financial and fiscal support policies, lower the threshold for farmers and migrant workers to start businesses, strengthen the construction of grass-roots employment service platforms and training bases, and strengthen entrepreneurship training for returning migrant workers [2]. Combined with the integrated development of rural primary, secondary, and tertiary industries and the construction of characteristic towns, a number of farmers' entrepreneurship parks and returning migrant workers' entrepreneurial parks will be built according to local conditions. These relatively slow development projects require a longer-term plan to enhance the existing social infrastructure, provide subsidies and support, and promote opportunities to younger generations, professionals, and local businesses.

12.5 Pathways to Carbon Neutrality in the Beibu Gulf City Cluster

The BGCC's carbon neutrality plan has not developed much in recent years. But since the 11th of July 2022, when the Guangxi Beibu Gulf Bank released a carbon neutral bond, we could see plans are progressing in a new way. Guangxi Beibu Gulf Bank issued the "*first carbon neutral-themed green financial bond for city commercial banks in the country in the inter-bank bond market, specifically supporting the construction of green projects with significant carbon emission reduction benefits*". China Daily [14]. As per reports, this is a significant "*innovative measure for the bank to implement the concept of green and sustainable development, as well as contributes to the country's dual carbon strategy of reaching carbon peak by 2030 and achieving carbon neutrality by 2060*" [14]. Following from this initiative, the expectations are high in line with the carbon neutrality targets, to achieve "*an annual reduction of 217,500 metric tons of carbon dioxide, 105,800 tons of standard coal, 55.52 tons of SO2, 62.11 tons of NOx, as well as 11.10 tons of smoke and dust, achieving significant carbon reduction*" [14]. Apart from the so-called green financial bonds, ongoing work is focused on dramatic marine environment change at the regional level [62], which requires further attention for ecological enhancement measures, ecosystem services restoration, and harmonious development. In the ports, a major plan proposed by the [18], is on green development strategies. In 2021, there were 72 special projects specifically focused on environmental protection and renovation. A considerable amount of investment of 430 million Yuan in a year helps to create at least a few green development demonstration sites and companies before 2025. These are considered part of the proposed systematic steps to achieve green development at multiple scales and integrate it into the ports' construction, planning, design, and operation.

From the ongoing studies, we note the importance of environmental protection and renovation is related to the status of economic development at smaller scales. As a bottleneck in the BGCC, we see the plan for economic zone development could nurture opportunities for carbon neutrality targets. The first proposed plan is to achieve carbon neutral operations by 2030 [1], using advanced technologies and innovative mechanisms in order to minimise the ecological footprint of new development projects. Key focus areas are aligned with carbon neutral targets. They propose data centers, carbon offsets, smart transportation development, low carbon and smart office buildings [13], AI cloud integration, and supply chain development [1]. These are just a few initiatives or interventions under development by Baidu in the BGCC. As a transitional process [11], we note the importance of industrial cluster development, industrial structuring, and further consideration of environmental values in the BGCC. This region is important due to its strategic location, but its current and future developments have to be more considerate than the development patterns/trends of the past few years.

References

1. Baidu Inc. (2021) Baidu announces goal to achieve carbon neutral operations by 2030. Available from: https://www.prnewswire.com/news-releases/baidu-announces-goal-to-achieve-carbon-neutral-operations-by-2030-301317231.html
2. Baike (2022) Beibu Gulf city group. <北部湾城市群>. Available from: https://baike.so.com/doc/5512008-5747771.html
3. Bing C (2022) The belt and road initiative: multilingual opportunities and linguistic challenges for Guangxi, China. Contemp Soc Sci 2022(2):8
4. Butollo F (2015) Growing against the odds: government agency and strategic recoupling as sources of competitiveness in the garment industry of the Pearl River Delta. Camb J Reg Econ Soc 8(3):521–536
5. Cao CM (2011) The analysis of Guangxi's cultural and creative industries development. In: 2011 IEEE 18th international conference on industrial engineering and engineering management. IEEE, pp 2110–2113
6. Chen F, Lin J, Qian B, Wu Z, Huang P, Chen K, Li T, Cai M (2018) Geochemical assessment and spatial analysis of heavy metals in the surface sediments in the eastern Beibu Gulf: a reflection on the industrial development of the south China coast. Int J Environ Res Public Health 15(3):496

7. Chen J (2018) Horizontal integration in the port industry in China challenges and outlook. Available from: https://commons.wmu.se/cgi/viewcontent.cgi?article=2496&context=all_dissertations
8. Chen J, Fei Y, Lee PTW, Tao X (2019) Overseas port investment policy for China's central and local governments in the belt and road Initiative. J Contemp China 28(116):196–215
9. Chen X (2007) A tale of two regions in China: rapid economic development and slow industrial upgrading in the Pearl River and the Yangtze River Deltas. Int J Comp Sociol 48(2–3):167–201
10. Cheshmehzangi A (2020) Towards urban energy management: a brief review of China's recent transition in urban energy demand. Low Carbon Econ 11(01):1
11. Cheshmehzangi A, Chen H (2021) China's sustainability transitions: low carbon and climate-resilient plan for carbon neutral 2060. Springer Nature, Singapore
12. Cheshmehzangi A, Dawodu A, Sharifi A (2021) Sustainable urbanism in China. Routledge, New York
13. Cheshmehzangi A, Li H, Yang X (2018) Low carbon and climate resilient urban development in China. ADB Report, developed for the National Development Reform Committee (NDRC), Beijing, China
14. China Daily (2022) Beibu Gulf Bank releases carbon neutral bonds. Available from: http://guangxi.chinadaily.com.cn/ftz/2022-07/14/c_788581.htm
15. Fang C, Yu D (2020a) The spatial pattern of selecting and developing China's urban agglomerations. In: China's urban agglomerations. Springer, Singapore, pp 65–126
16. Fang C, Yu D (2020) China's urban agglomerations. Springer, Cham, pp 1–32
17. Gong B, Cao H, Peng C, Perčulija V, Tong G, Fang H, Wei X, Ouyang S (2019) High-throughput sequencing and analysis of microbial communities in the mangrove swamps along the coast of Beibu Gulf in Guangxi, China. Sci Rep 9(1):1–10
18. Guangxi Government (2021) Beibu Gulf ports focuses on green development. Available from: http://en.gxzf.gov.cn/2021-08/30/c_655786.htm
19. Han H (2019) The reference of China's regional cooperation and development experience to the cooperation of Huaihai economic zone. Acad J Humanities Soc Sci 2(2)
20. Hong Y (2018) Regional Development in China: Xi Jinping's agenda and the challenges. China: Int J 16(3):179–199
21. Hosokawa D (2009) Pan-Beibu Gulf economic cooperation. In: China's new initiative in cooperation with ASEAN. Osaka Keidai Ronshu, vol 60, No. 2, pp 67–78. Available from: https://www.i-repository.net/il/user_contents/02/G0000031Repository/repository/keidaironshu_060_002_067-078.pdf
22. Huang GY (2017) Suggestions on development strategies and policies of national parks in Qinzhou under strategy of" One Belt and One Road". In: 2017 3rd international conference on economics, social science, arts, education and management engineering (ESSAEME 2017). Atlantis Press
23. Huang Z (2012) Guangxi's economic gains from cooperation with neighboring countries in the greater Mekong Sub-Region (GMS). J WTO China 2:93
24. Jiang F, Du C, Xiao R (2018) The Function and role of rural revitalization in the development of urban agglomeration. In: The international conference on economics, finance, business, and development (ICEFBD 2018). Available from: https://webofproceedings.org/proceedings_series/ECOM/ICEFBD%202018/ICEFBD043.pdf
25. Jiang Y, Qiao G, Lu J (2020) Impacts of the new international land–sea trade corridor on the freight transport structure in China, central Asia, the ASEAN countries and the EU. Res Transp Bus Manag 35:100419
26. Lan Q (2011) An analysis of the development of logistics in Guangxi Beibu Gulf economic zone. In: Advanced materials research, vol 225. Trans Tech Publications Ltd., pp 178–181
27. Lao Q, Liu G, Zhou X, Chen F, Zhang S (2021) Sources of polychlorinated biphenyls (PCBs) and dichlorodiphenyltrichloroethanes (DDTs) found in surface sediment from coastal areas of Beibu Gulf: a reflection on shipping activities and coastal industries. Mar Pollut Bull 167:112318

28. Li F, Xing W, Su M, Xu J (2021) The evolution of China's marine economic policy and the labor productivity growth momentum of marine economy and its three economic industries. Mar Policy 134:104777
29. Li L, Zhao K, Wang X, Zhao S, Liu X, Li W (2022) Spatio-temporal evolution and driving mechanism of urbanization in small cities: case study from Guangxi. Land 11(3):415
30. Li M, Kwa CG (2011) China-ASEAN sub-regional cooperation: progress, problems and prospect. World Scientific
31. Li X, Pathak S (2021) Feasibility study of Beibu Gulf Shipping exchange project. Sci, Technol, Soc Sci Procedia 2021(1):acm006–acm006
32. Ling Z, Jiang W, Liao C, Li Y, Ling Y, Peng K, Deng Y (2022) Evaluation of production–living–ecological functions in support of SDG Target 11. a: case study of the Guangxi Beibu Gulf Urban Agglomeration, China. Diversity 14(6):469
33. Liu J (2018) Development of regional logistics in the belt and road. In: Contemporary logistics in China. Springer, Singapore, pp 205–221
34. Liu P (2020) Marine ecological construction and economic coordination in Beibu Gulf city cluster of China. J Coast Res 109(SI):19–26
35. Liu S, Liu Y (2022) Paths for Zhanjiang's participation in the construction of the New Western land-sea corridor. Ind Eng Innov Manag 5(1):64–70
36. Liu Z, Schindler S, Liu W (2020) Demystifying Chinese overseas investment in infrastructure: port development, the Belt and Road Initiative and regional development. J Transp Geogr 87:102812
37. Ma Q, Jia P, She X, Haralambides H, Kuang H (2021) Port integration and regional economic development: lessons from China. Transp Policy 110:430–439
38. National Standard Full Text Disclosure System (2017) Beibu Gulf urban agglomeration development plan. <北部湾城市群发展规划>. Available from: http://www.77zjs.com/4312.html
39. Nie Y, Chen N, Chen Z (2021) Analysis of the development situation of private higher education in the Beibu Gulf city cluster. In: 2021 2nd asia-pacific conference on image processing, electronics and computers, pp 594–599
40. Qiao P, Li Y (2021) Research on performance evaluation of Beibu Gulf Port development: an empirical study of 11 years of data. In: E3S web of conferences, vol 275. EDP Sciences, p 02018
41. Ou M, Lai X, Gong J (2022) Territorial pattern evolution and its comprehensive carrying capacity evaluation in the coastal area of Beibu Gulf, China. Int J Environ Res Public Health 19(17):10469
42. Peng D, Li R, Shen C, Wong Z (2022) Industrial agglomeration, urban characteristics, and economic growth quality: the case of knowledge-intensive business services. Int Rev Econ Financ 81:18–28
43. Schmalz S, Sommer B, Xu H (2018) The Yue Yuen strike: Industrial transformation and labour unrest in the Pearl River Delta. In: Chinese labour in the global economy. Routledge, pp 107–119
44. Song J (2009) Estimate the rationality of spatial structure of Beibu Gulf (Guangxi) economic zone and analyze the agglomeration trend. Int J Bus Manag 4(1):132–137
45. Song X, Wu S, Xu X (2019) New trends of regional development since the 18th CPC national congress. In: The great change in the regional economy of china under the new normal. Palgrave Macmillan, Singapore, pp 161–191
46. Tao Y, Li F, Crittenden J, Lu Z, Ou W, Song Y (2019) Measuring urban environmental sustainability performance in China: a multi-scale comparison among different cities, urban clusters, and geographic regions. Cities 94:200–210
47. Wang J, Cheng J, Qin N, Chen Q (2017) The spatial network of Beibu Gulf urban agglomeration and its challenges for coordinated integral development. In: Jinan forum on geography and ecological sustainability Guangzhou, China. Springer, Cham, pp 307–324
48. Wang MX, Zhao HH, Cui JX, Fan D, Lv B, Wang G, Li ZH, Zhou GJ (2018) Evaluating green development level of nine cities within the Pearl River Delta, China. J Clean Prod 174:315–323

49. Wei H, Nian M, Li L (2020) China's strategies and policies for regional development during the period of the 14th five-year plan. Chin J Urban Environ Stud 8(02):2050008
50. Wong Z, Chen A, Shen C, Wu D (2022) Fiscal policy and the development of green transportation infrastructure: the case of China's high-speed railways. Econ Chang Restruct 1–35
51. Xiong X, Tang Y (2014a) Strategic development of Beibu Gulf economic zone of Guangxi: from the perspective of low carbon economy. In: Proceedings of the seventh international conference on management science and engineering management. Springer, Berlin, pp 1251–1260
52. Xiong X, Tang Y (2014b) the industrial development of Beibu Gulf economic zone of Guangxi: issues and measures. In: Proceedings of the eighth international conference on management science and engineering management. Springer, Berlin, Heidelberg, pp 1257–1267
53. Yang J, Luo M, Wu X (2022) How port integration policy affects total factor productivity in Chinese ports: empirical evidence from 2005 to 2018. J Transp Econ Policy (JTEP) 56(1):79–102
54. Yu H, Lim WX (2014) The Guangxi Beibu Gulf economic zone and China's maritime silk road in the 21st century. East Asian Policy 6(03):16–29
55. Yu J, Zhou K, Yang S (2019) Land use efficiency and influencing factors of urban agglomerations in China. Land Use Policy 88:104143
56. Zhang H (2021) Analysis of the influence of Beibu Gulf port development on regional economy. Master Dissertation at the World Maritime University, Shanghai, China. Available from: https://commons.wmu.se/cgi/viewcontent.cgi?article=2626&context=all_dissertations
57. Zhang Q, Jiang W (2011) Research on the development of tourism in Beibu Gulf from the perspective of comparing Gulfs worldwide. In: 2011 2nd International conference on artificial intelligence, management science and electronic commerce (AIMSEC). IEEE, pp 1432–1438
58. Zhang X (2020) The belt and road: building a community of shared interests together. In: The difficult task of peace. Palgrave Macmillan, Cham, pp 203–224
59. Zhao D, Xiao M, Huang C, Liang Y, Yang Z (2021) Land use scenario simulation and ecosystem service management for different regional development models of the Beibu Gulf Area, China. Remote Sens 13(16):3161
60. Zhao H (2019) China-Japan compete for infrastructure investment in Southeast Asia: geopolitical rivalry or healthy competition? J Contemp China 28(118):558–574
61. Zhong C, Yang Z, Jiang W, Hu B, Hou Q, Yu T, Li J (2016) Ecological geochemical assessment and source identification of trace elements in atmospheric deposition of an emerging industrial area: Beibu Gulf economic zone. Sci Total Environ 573:1519–1526
62. Zhou G, Cao X, Xia J, Wang S, Song Z (2022) A dramatic marine environment change in the Beibu Gulf of the South China sea around 3.2 kyr BP. Lithosphere Special 9:2632112. https://doi.org/10.2113/2022/2632112.
63. Zhu Q, Sun C, Zhao L (2021) Effect of the marine system on the pressure of the food–energy–water nexus in the coastal regions of China. J Clean Prod 319:128753

Chapter 13
Megalopolis in Central Yunnan: From Local Development to Regional Industrial Clusters, Tourism, and Sustainable Production

13.1 A Brief Introduction to the Central Yunnan City Cluster

The urban agglomeration in central Yunnan consists of seven counties and cities, including cities of Kunming, Qujing, Yuxi, Chuxiong Yi Autonomous Prefecture, and Mengzi. It also includes Gejiu, Jianshui, Kaiyuan, Mile, Luxi, and Shiping counties in the north of Honghe Hani and Yi autonomous prefecture. The Central Yunnan City Cluster (CYCC) covers a total area of 114,600 km^2 [35]. In 2018, it had an estimated population of 21.27 million (end of 2018) and a total GDP of 1.02 trillion yuan.

The CYCC is one of the 19 urban agglomerations that the state has focused on cultivating. It is the last of the city clusters we map in this book. There is either little information or no carbon neutrality (or both) for the remaining eight city clusters, which are generally of lower level importance at the national level. For the CYCC, we note several significant factors. It is an integral part of the "two horizontal and three vertical" urbanization strategy [35], the key zone of western development, and the critical growth pole of China relying on the Yangtze River to build a new economic support belt [13]. With the implementation of the Belt and Road Initiative (BRI), the regional advantage of opening up to South and Southeast Asia of the central Yunnan urban agglomeration is further highlighted [11], ushering in a period of strategic opportunity for rapid development.

As the core of Yunnan, the CYCC leads the development of the whole province. It plays a decisive role in integrating and serving the national development strategy, accelerating the pattern of Yunnan's opening to the outside world, and implementing the new urbanization development ideas [8]. It is also aligned with the requirements of the country and Yunnan Province. The CYCC development takes the new urbanisation road with Yunnan characteristics, fully realise the 13 and 14th five-year Plans' goals, the "two hundred years" goal, and so on. Due to its location neighbouring countries in the ASEAN region, the CYCC plays a significant role in China's international cooperation, transportation development, infrastructural projects, communications, etc.

A. Cheshmehzangi and T. Tang, *China's City Cluster Development in the Race to Carbon Neutrality*, Urban Sustainability, https://doi.org/10.1007/978-981-19-7673-5_13

13.2 A Recent Overview of the Central Yunnan City Cluster Development Plan

At first, we look at the CYCC's planning process and provide an overview of its development since 2009. In October 2009, the Development and Reform Commission of Yunnan Province published the Coordinated Development Plan of Urban Economic Circle in Central Yunnan (2009–2020) for public comment. On May 28, 2010, the Department of Housing and Urban–Rural Development of Yunnan Province and other provincial departments, as well as the relevant states and cities in the Joint Review Mission, conducted a review of the "Revision of the Planning of Urban Agglomeration in Central Yunnan 2009–2030" [35]. On May 27, 2011, the Yunnan Provincial People's Government officially approved the Central Yunnan urban agglomeration planning (2009–2030). On August 3 of the same year, Yunnan held a press conference [35] to officially announce the planning of the CYCC.

On December 23, 2016, according to the work deployment of the Yunnan Provincial Party Committee and Yunnan provincial people's government, the Department of Housing and Urban–Rural Development of Yunnan province organised and completed the compilation of the "Plan for the Central Dian City Cluster (2016–2049)" [35]. In accordance with the Law of the People's Republic of China on Urban–Rural Planning and the Regulations of Yunnan Province on Urban and Rural Planning, "The Planning of Central Yunnan Urban Agglomeration (2016–2049)" (Public Draft) is hereby published online [13] to solicit opinions and suggestions from all sectors of society.

13.3 Summary of the Development and Industrial Structure in the Central Yunnan City Cluster

The middle area of Yunnan province is relatively rich in land resources with relatively flat terrain. The climate is exceptionally conducive to people's physical and mental health, especially suitable for human habitation [35]. The mineral resources reserve is large. The economic value is high. At the regional level, the resources are extremely abundant, concentrated in the vast majority of phosphorus, copper, iron, lead, coal, and other minerals in Yunnan [12]. Kunming is one of the first national historical and cultural cities with rich tourism and cultural resources. Chuxiong prefecture is one of China's only two Yi autonomous prefectures [13, 36]. It has several national scenic spots and tourist resorts. It is an important tourist resort in South and Southeast Asia. It is the main producing area of grain, flue-cured tobacco, vegetable, flower, animal husbandry, and so on. The overall level of the ecological environment is suitable [4], and the forest coverage rate is more than 50% of the region. Coupled with suitable climatic conditions conducive to animal and plant growth and ecological environment recovery [31], it is evident that the province's ecological environment carrying capacity is robust.

The present infrastructure situation has a slightly different story from what has been perceived over the years. The construction of traffic facilities in the CYCC has achieved outstanding results. The bottleneck constraint of traffic on development has been basically eased, and the modern comprehensive traffic network is taking Kunming as the center has taken shape [29]. The low utilisation rate of water resources [32], the outstanding problems of engineering water shortage, and the outstanding contradiction between supply and demand have become one of the biggest bottlenecks restricting the economic and social development in the central part of Yunnan [12]. Over the years, the level of energy security has markedly improved [23], the development of power sources has been advancing in an orderly manner, and the main power grid structure has been continuously improved. It is noted that breakthroughs have been made in the construction of oil and gas pipelines [35], and the China-Myanmar natural gas pipeline has been completed and put into operation [13]. The ability to support the Internet infrastructure continued to improve. This led to the development of the Internet network structure, which took place in recent years.

The present situation of industrial economic development shows that the economy of central Yunnan urban agglomeration grows rapidly based on the function of central city opportunities. The economic development level and industrial structure of each city in this urban agglomeration differ greatly. Thus, the leading industries of each state and city in the urban agglomeration are convergent. However, the domestic and international competitiveness is not obvious.

Currently, there is a general population and urbanisation development situation at the regional level. The total population growth is stable, but the population growth varies greatly among regions [36]. The ongoing rapid urbanisation has its own issues. For instance, the overall level of the region is not of high significance, and there is a lack of intermediate-scale cities, showing the characteristics of high priority. The current situation of resources and environment development indicates the problem of decreasing the self-sufficiency of resources and energy. This is a prominent and growing issue, including the shortage of effective water resources that restricts economic and social development [36]. The region has the most advantageous land resources in Yunnan, but the contradiction in land use is still outstanding [35]. The ecological environment is an important support to realise sustainable development in the middle of Yunnan Province, but the environmental pollution intensifies, and the environmental protection pressure is heavy.

13.3.1 An Overview of the Regional Contradictions

The overall level of development is not high in the CYCC region. The urban agglomeration's regional GDP and per capita GDP in central Yunnan ranked 17th among the 19 urban agglomerations in China, with an obvious development gap [35]. The modern industrial system is not perfect, the export-oriented industry is not fully developed, the export product level is low, the scale is small, and the industrial development homogeneity phenomenon is relatively prominent [22, 38]. These are some of the main issues in this urban agglomeration. The overall quality of urbanisation is not high, and the spatial structure of cities and towns needs to be optimised [36]. As it stands, the population gathering capacity of central cities is insufficient, and the radiation and driving effect is not strong.

At the regional level, internal and external openness is not high. The service level of the external channel is low, the transportation cost is high, the unblocked guarantee is insufficient, and the comprehensive transportation efficiency is not high. The level of cooperation with neighbouring regions needs to be improved, showing the characteristics of scattered, small, and weak development [6, 18, 25]. Thus, we can verify that regional development is not yet formed via extensive cooperation with South and Southeast Asia countries, especially Myanmar, Laos, and Vietnam [35]. More importantly, Kunming, the central city in this cluster, lags behind in the development of export-oriented service industries such as finance, trade, logistics, and information.

Furthermore, it is noted that infrastructure network support needs to be improved. There is insufficient hardware connectivity and joint construction and sharing of infrastructure. The degree of the comprehensive transportation network is low, and the connection and communication with the transportation infrastructure and logistics system of neighbouring countries and regions are insufficient. As per the governmental reports, the aviation network system for South and Southeast Asia is not perfect, and the traffic points and flight frequency need to be further increased. Unfortunately, the COVID-19 pandemic has also affected the regional transportation and communication links, particularly with the neighbouring countries of Southeast Asia. The construction of water conservancy facilities lags behind, and the water shortage situation remains unchanged [35]. Information platforms such as "One Mobile Phone Travel in Yunnan" and "One Mobile Phone Service" need further improvement and upgrades. In light of these challenges, the information infrastructure is insufficient to support regional economic and social development.

On top of these regional contradictions, the mechanism for coordinated regional development is not yet sound [35]. The working mechanism of regional joint meetings has not yet been put into regular operation. In light of such weak cooperation, local interests dominate, and market segmentation still exists [36]. Administrative barriers have not yet been removed, thus, factor flow is not smooth [12]. The impacts are severe as the regional unified market and credit system construction is not yet completed and lags behind [33]. There are also significant issues with the urban agglomeration integration development cost-sharing [16, 17] and benefit-sharing mechanism, which have not been established.

13.4 The Cluster's Sustainability Agenda: Opportunities for Sustainable Development

The state has made it clear that Yunnan should take the initiative to serve and integrate itself into the national development strategy and blaze a path of leapfrog development [35]. As part of this plan, efforts will be made to build Yunnan into a demonstration zone of national unity and progress, a pacesetter in ecological civilization construction [5], and a radiation center for South and Southeast Asia. The government also points out a new direction for developing central Yunnan urban agglomeration [35]. The profound adjustment of the international and regional economic structure, the construction of the Bangladesh-India-Myanmar Economic Corridor, the China-Indo-China Peninsula economic corridor, and the Yangtze River economic belt provide new opportunities for the opening up of the urban agglomeration in central Yunnan [35, 36]. The simultaneous development of new-type industrialisation, informatisation, urbanisation, agricultural modernisation, and the overall deepening of reform have injected new impetus into the economic and social development of the CYCC [7, 24]. The implementation of the western development strategy provides a new guarantee for the leap-forward development of central Yunnan urban agglomeration [35]. The overall acceleration of the construction of cross-border and trans-regional infrastructure systems and the industrial orientation of "two types and three changes" provide new support for the leapfrog development of the CYCC.

A major emphasis is on deepening reform and innovation development [35]. In this regard, the government plans to strengthen institutional innovation, remove administrative barriers, improve a unified market system that promotes the free flow of factors, and improve the mechanisms for integrated development and two-way opening up and coordination [36]. These strategies enhance the regional ability to innovate in science and technology and strengthen the transformation of scientific and technological achievements [12]. Thus, the central proposal is to jointly build a technological innovation chain and a regional collaborative innovation system. This approach will help build a modern economic system and strengthen international and domestic competitiveness.

In addition, the overall planning and coordinated development [35] require further attention. According to the carrying capacity of resources and environment and the suitability of land space development, the government and involved stakeholders should optimise the development pattern of regional land space and make overall plans for economic and social development, spatial population distribution, ecological and environmental protection, infrastructure construction [36], and domestic and external opening up [28]. The associated plan is to identify the functional positioning of cities at all levels, promote regional coordinated development, and promote the complementary advantages of the state and city through differentiated development mechanisms.

Lastly, a significant direction is to improve supporting facilities and share development [35]. In his regard, the government continues to put the people at the center, focus on overall human development, and improve public service systems in education, culture, medical care, old-age care, childcare, sports, and social security. The government targets to make overall plans for the layout of major regional infrastructure and realise the co-construction and sharing of regional infrastructure [3, 36]. These strategies will gradually improve the equalisation of urban basic public services [35] and promote people in central Yunnan to share the fruits of development.

13.4.1 Sustainable Development Plan from the Environmental Perspective

The first sustainable development plan for the CYCC is to deepen cross-regional water pollution prevention and treatment. This strategy is planned through the implementation of the District Watershed Governance Strategy. The total nitrogen and phosphorus pollutants, which have a prominent influence on the environmental water quality, were selected and brought into the river basin's total emission control index system [35]. In the headwaters of rivers, drinking water source protection zones, and their upper reaches, the development of high-risk and high-polluting industries is strictly prohibited. Consequently, the low-level expansion and redundant construction of industries with high energy consumption and high emissions are strictly controlled, and the chemical industry and other industries are shut down and closed [35]. A mechanism for joint prevention and control of the ecological environment in cross-border cross-sections and compensation for ecological protection in river basins should be established [36]. Emphasis will be given to the comprehensive improvement of the water environment of the Jinsha River, Nanpan Jiang, Honghe, Niulan, and Huangni Rivers, as well as the Pudu, Lu, Dianxi, Qu, and Luzhi rivers that cross the city cluster. More importantly, the government plans to strengthen the construction of urban and rural sewage networks along the BRI through green development opportunities [11]. In doing so, the ongoing and forthcoming projects promote the upgrading of sewage treatment facilities and the utilisation of resources in line with local conditions and do an excellent job in collecting and treating domestic sewage [35]. To achieve this, the government pledges to make overall efforts to improve the rural environment [35], strengthen control of agricultural non-point source pollution, promote the reduction of pesticides and chemical fertilizers, and develop eco-recycling agriculture.

The second overarching plan is to deepen the prevention and control of groundwater pollution. The government continues to promote the urban centralised groundwater drinking water supply area, chemical enterprises, gas stations, landfills, and hazardous waste disposal sites around the basic environmental status of groundwater investigation [35]. Such strategies would be carried out by investigating and detecting seepage prevention, implementing anti-seepage measures for serious leakage, and formulating measures to promote anti-seepage work at key pollution sources comprehensively. So far, the government has reported that steady progress has been made in controlling the risks of groundwater pollution in critical areas and remediation of groundwater contamination in key industries at the regional level.

The third major plan at the regional level is to coordinate pollution control of the six lakes. This strategy adhered to the "one lake, one policy" approach to governance, the implementation of accurate governance, and systematic remediation [35]. The government pledges to fulfill the responsibilities of lake governors, improve the quality of legislation for the protection of the six lakes, push forward the implementation of major pollutant reduction projects, and improve the total emission control system for major pollutants [35]. A major focus aligned with this plan is on establishing the lake basin's ecological security pattern and promoting the "four retreat and three return" (retreat people, farmland, pond, lake, water, and wetland). This approach will improve the efficiency of water use and restore the ecological functions of the basin. The Fuxian Lake with good water quality [1, 20, 30], and Yangzonghai with better water quality, continue to strengthen pollution monitoring and risk prevention to comprehensively improve the environmental water quality [36]. To further eliminate the risk of arsenic pollution in Yangzonghai, comprehensive prevention and control of surrounding pollution are essential. This approach would strengthen the dual pollution control in the development zone and rural life. The Dianchi, Qilu, Xingyun, and Yilong Lakes, which are heavily polluted, should be brought under comprehensive control, with a view to improving the carrying capacity of water resources and the quality of the water environment in the lakes. Thus, it is urgent to strengthen the construction of sewage treatment facilities, support pipe networks in cities and towns along river valleys and improve the effect of pollution interception and control [35]. Dianchi Lake implements the strictest control system via strictly controlling tourism facilities [2] and real estate projects along the lakeshore [35]. Other associated strategies are to improve the precision pollution control system of Dianchi Lake, strengthen the exemplary management of the whole basin [35], and improve the efficiency of the sewage treatment system and the sewage interception system.

The fourth sustainable development strategy under the regional sustainability agenda is to strengthen the prevention and control of air pollution. Yunnan Province has already implemented a three-year campaign to win the blue sky defense war, carried out a comprehensive fine particulate matter (PM2.5) and ozone monitoring, and significantly reduced the total amount of major air pollutants [35, 36]. The government continues to improve pollution control in key industries such as iron and steel, petrochemical, chemical, nonferrous metals, and building materials, and advance in an orderly manner the transformation of steel enterprises with ultra-low emissions. The aim is then to strengthen pollution prevention and control from mobile sources [35], such as motor vehicles, non-road mobile machinery, and ships, with diesel trucks as the focus.

The fifth overarching plan is to improve the sound environment quality, which requires rational distribution of urban construction projects to avoid noise nuisance [35]. This strategy is proposed to carry out the division and adjustment of the acoustic environment functional areas, implement the requirements of the functional areas in the urban construction, and develop the management of the quality of the urban acoustic environment [35]. Expected outcomes in the coming years are to improve the urban environmental noise monitoring system and strengthen the management of key noise pollution sources.

The sixth main regional sustainable development plan is to improve soil environmental quality. In this regard, the government plans to strengthen the protection and safe utilisation of agricultural land, implement classified management of agricultural lands, and ensure the quality and safety of agricultural products [35, 36]. These strategies would strengthen the management, control, and remediation of soil pollution risks for construction land. In doing so, the government aims to establish a directory of land pollution risk management and remediation and strengthen access management of land for construction according to law. These will be associated with better planning to supervise and urge the key supervision units of soil pollution [35] to fulfill their obligations of soil contamination prevention and control according to law.

The last overarching sustainable development plan promotes the joint prevention and control of solid waste, hazardous waste heavy metal pollution [35]. This plan helps to strengthen the prevention and control of industrial solid waste pollution, reduce the generation of solid waste from the source of production, promote the recycling of solid wastes and the development of a circular economy, and improve the capacity and level of harmless disposal of solid refuse [35]. Accordingly, the government aims to strengthen pollution prevention and control of large quantities of solid waste and comprehensively investigate and rectify industrial solid waste storage sites [35]. These plans will help strengthen the management of hazardous waste production units and business units from the source to eliminate the illegal transfer of hazardous wastes [35]. Rational allocation of hazardous waste disposal capacity is required further improvement, which needs to be aligned with the overarching plan. The associated strategies are also focused on strengthening pollution prevention and control in heavy metal industries, strengthening environmental supervision of enterprises in heavy metals industries, and promoting enterprises to achieve comprehensive emission standards [35]. In areas where mineral resources development activities are concentrated, special emission limits for major heavy metal pollutants shall be strictly enforced to prevent mineral resources exploitation from polluting the soil.

13.4.2 Sustainable Development Plan from the Social Perspective

The correlation between environmental and social development plans in the CYCC is evident in how the plans are strategised for the coming years and decades. The first is to promote the green development of cities and towns [19] in the region. This will be directed via strict planning and management, rational distribution of ecological land, and appropriate expansion of urban ecological space. The government aims to carry out ecological restoration and urban rehabilitation, promote urban renewal and transformation, strengthen the urban design, and enhance urban characteristics and vitality [15]. This overarching plan will also coordinate the development of underground and above-ground space, steadily advance the construction of urban underground utility corridors, and establish and improve a comprehensive urban disaster prevention system [35]. The associated strategies are to promote the development of cities and towns from the extension of the expansion to enhance the connotation of change [26]. Also, this will help develop green energy, promote green buildings, and build a green transportation system [35]. The expected outcomes or impacts are the formation of clean water, fresh air, clean soil, health, and livable environment.

The second plan under the social sustainability agenda of the CYCC is to promote the construction of beautiful and livable villages [35]. According to this plan, the government aims to implement the "Seven Actions to Build Beautiful and Livable Villages" and strengthen farmland protection, ecological conservation, infrastructure construction, and industrial development [35]. Combined with the actual characteristics of villages, according to the village policy, this strategy is in place to strengthen the planning and construction management of villages and towns, and adhere to the renovation of rural human settlements. Associated with this plan, the government seeks to build beautiful urban and rural areas. The strategy is to promote the coordinated development of new-type urbanization and new rural construction with people at the core. The government pledges to promote intensive and efficient production space, moderate living space, and beautiful ecological space to form a rational structure of production, living and ecological space, and build an integrated urban and rural development pattern. One of the main mottos for this plan is literally translated to "*Let the residents see the mountains, see the water, and remember nostalgia*" [35]. To achieve these, the government has made efforts to make overall plans for cities and towns, improve municipal infrastructure, promote a balanced allocation of public resources between urban and rural areas, and promote the construction of beautiful counties and beautiful villages. These will help the social development of rural areas and townships and provide important support for enhancing urban–rural development at the provincial level [35]. Aligned with this plan, the secondary strategies are to protect and carry forward the excellent traditional culture in central Yunnan, continue the historical context, and inherit the national culture.

One of the key areas of focus at the regional level is to build a green economy. To achieve this relatively vague plan, the government should at least establish and improve an ecological economic system in which industry is ecologically oriented and ecologically industrialised [21, 39]. Thus, they pledge to adjust the economic structure and energy structure, optimise the development pattern of territorial space, adjust the industrial layout of regional river basins, and promote the development of energy conservation and environmental protection industries, clean production industries, and clean energy industries [35]. The development of the circular economy also affects the promotion of recycling park transformation. Also, it would help develop ecological economy adapted to the environmental characteristics of mountainous and karst areas [35]. In this regard, the government has proposed to implement green financial policies such as green credit, green bonds, and green insurance and establish eco-industrial clusters with local characteristics. Aligned with the above-mentioned sustainable development plans, the government aims to advocate a low-carbon lifestyle [10]. In doing so, the aim is to establish a sound ecological, cultural system with ecological values as the criterion. Such an approach would help foster ecological culture, guide green consumption, encourage low-carbon travel, advocate simple moderation, and achieve a green and low-carbon lifestyle [35]. More importantly, and also linked to achieving carbon neutrality goals, the government has already initiated plans to promote the individual low carbon plan, carry out low-carbon family action, and promote low-carbon community construction.

13.4.3 Sustainable Development Plan from the Economic Perspective

One of the very few larger-scale economic development plans is to create the source of strategic emerging industries [35]. The government aims to optimise the distribution of industries, promote the effective concentration and supply of talent, technology, and capital, and foster new drivers of industrial development, thus, becoming a critical leading engine for economic transformation and upgrading [35]. Focused on the new generation of information technology and manufacturing technology, it is essential for the government to consider deep integration of innovation, vigorously develop intelligent manufacturing technology, and upgrade their intelligent manufacturing equipment. In the biomedicine and health industry, the aim is to lay out the industrial chain and innovation chain around the innovation direction of drugs for serious diseases, medical devices, and personalised treatment. The focus is on building biomedicines and large health industry clusters such as biomedics and biomedical engineering. The government plans to implement the cloud-on-cloud action plan, improve data transmission speed and capacity, and promote the in-depth cross-border integration and interactive development of the Internet, big data, and other technologies with various industries [35]. These mechanisms are set to forward the layout of cutting-edge new materials research and development and build new materials industry systems.

Another major economic development plan of the CYCC is to build a modern service industry gathering area. Implementing the service economy dual or double plan is aligned with the regional level carbon neutrality targets. The government continuously encourages coordinated financial development and vigorously develops efficient, safe, green, inclusive, open, and innovative modern financial services [35]. The associated plans are to develop efficient modern logistics, establish a modern logistics service system, build a multi-modal intelligent logistics park, and promote the construction of an intelligent logistics network in the CYCC [35]. Subsequently, the government aims to improve the service level of scientific and technological innovation, speed up the development of science and technology R&D and entrepreneurial services, and build a service system for scientific innovation. They also plan to actively build scientific innovation service centers, cultivate market-oriented scientific innovation services, and brand service institutions [35]. To achieve these, the plan is to jointly build cultural and tourism brands and establish a linkage mechanism between tourism destinations. The government also proposed to jointly create high-quality cultural tourism products and well-known tourism brands at home and abroad, form a diversified, distinctive, and differentiated tourism product system [35], and implement tourism information exchange [27], and promote the development of cultural and tourism industry clusters.

Lastly, another major plan is to build demonstration zones for the transformation and upgrading of traditional industries. This plan helps promote greater integration of IT application and industrialisation, pursue green development, reshape the industrial, value, and market chains, transform and upgrade key industries, activate existing stocks and boost driving forces, and make new contributions to steady growth [35]. Focused on the construction of tobacco and supporting industrial clusters, the strategies are set to improve the level of modern tobacco agriculture, speed up the adjustment of product structure, and develop the tobacco-supporting industry [35]. As per governmental reports, it is strictly prohibited to increase smelting capacity in the metallurgical industry. The secondary strategies for this plan are to transform and upgrade traditional industries, promote new intelligent manufacturing models, improve the level of industry technology and equipment and the comprehensive utilisation of resources, and develop green and low-carbon energy-carrying industries [35]. The development of the petrochemical processing industry chain is also aligned with this overarching plan. In doing so, the government plans to develop fine chemicals vigorously, optimise the chemical fertilizer product mix, and do a better and stronger chlorination titanium dioxide industry [14, 37]. These indicators require the steady development of a new coal chemical industry [35], which would be important for the future development of the industrial formation, readjustment, and cluster development.

Furthermore, the government aims to speed up the adjustment of building materials industry structure, product mix, the development of new building materials and deep processing products, new wall materials, the stone industry, and the ceramic industry [35]. They will further advance supply-side structural reform in agriculture, vigorously develop modern agriculture with plateau characteristics, innovate agricultural development models, and develop diversified forms of agriculture and appropriately scaled-up operations [35]. These would help explore ways to establish production bases for export-oriented, high-quality, superior, and distinctive agricultural products with neighbouring countries and regions [35]. Subsequently, the plan is to actively participate in the construction of the cross-border economic belt of the agricultural industry, strengthen the ability of agricultural science and technology innovation, vigorously cultivate agricultural scientific and technological enterprises and high-tech enterprises, and develop and build high-tech agricultural industry.

13.5 Pathways to Carbon Neutrality in the Central Yunnan City Cluster

The mapped CYCC's sustainable development plans in the previous section help us picture its carbon neutrality targets aligned with the national-level objectives. As mentioned before, there is a diverse range of industries and industrial clusters in the CYCC. Thus, we see directions are aligned with each area's local capacities and status, such as the development of China's first zero-carbon food factory [34], which focuses on dairy products, accelerating the green development agenda, energy saving, and carbon reduction. However, regional low-carbon development could become slightly tricky due to differences across the province. Nonetheless, the CYCC has already performed well in dropping its carbon intensity, where they were ahead of its targets in 2020. The continuous efforts till 2025 and then meeting the 2030 carbon peak targets would enable regional development [9] to be aligned with low-carbon strategies and transitions.

As highlighted in the earlier sections, the CYCC's economic and social sustainability plans are closely linked. This means that the two targets should be integrated to achieve the carbon neutrality goals. This is particularly important for the regions and similar city clusters of the same level or category. Hence, despite the many environmental restoration and biodiversity protection plans (such as the post-2020 global biodiversity protection plan), the aims should provide financial support and infrastructural development for achieving carbon neutral goals. Such approaches should be strategised at the smaller scale of cities or counties and at the regional level, which may be the bottleneck for the CYCC. In sum, coordination is needed at the regional level to ensure cities are not left alone, urban–rural cooperation is enhanced, and the primary energy consumption is reduced. Lastly, we urge the next steps of the CYCC's development to be focused on its cross-provincial and international transportation and communication links, which may help nurture what would be a better circular economy development at the regional level.

References

1. Bai Y, Li C, Qin S, Xie N (2017) Exploration of the new-type urbanization paths for Yuxi three-lake ecological urban agglomeration. In: 2017 9th international economics, management and education technology conference (IEMETC 2017). Atlantis Press, pp 352–356
2. Cai Z, Ding L (2016) Research on the tourism industry transformation of the northwest region of Yunnan Province from the perspective of extension marketing theory. In: 2016 5th international conference on social science, education and humanities research (SSEHR 2016). Atlantis Press, pp 1266–1271
3. Chen J, Xu B, Tang W, Zou W, Yang W, Liang L, Zhan Z, Tang M, Bai F, Chen Y, Yang D (2021) Cloud computing based big data platform construction of university informatization teaching service in Yunnan Province under the background of community of destiny. In: MATEC web of conferences, vol 336. EDP Sciences, p 05015

4. Chen Y, Wang J, Kurbanov E, Thomas A, Sha J, Jiao Y, Zhou J (2022) Ecological security assessment at different spatial scales in central Yunnan Province, China. Plos one 17(6):e0270267
5. Chen Y, Wang J (2020) Ecological security early-warning in central Yunnan Province, China, based on the gray model. Ecol Ind 111:106000
6. Cheng G, Li S, Deng C (2011) Research on the agro-processing industry in Yunnan participating in international regional cooperation. In: 2011 international conference on management and service science. IEEE, pp 1–5
7. Cheshmehzangi A (2014) The urban and urbanism: China's new urbanization and approaches towards comprehensive development. Int J Interdiscip Environ Stud 8(3–4):1–12
8. Cheshmehzangi A (2016) China's New-Type Urbanisation Plan (NUP) and the foreseeing challenges for decarbonization of cities: a review. Energy Procedia 104:146–152
9. Cheshmehzangi A, Chen H (2021) China's sustainability transitions: low carbon and climate-resilient plan for carbon neutral 2060. Springer Nature, Singapore
10. Cheshmehzangi A, Li H, Yang X (2018) Low carbon and climate resilient urban development in China. ADB Report, developed for the National Development Reform Committee (NDRC), Beijing, China
11. Cheshmehzangi A, Xie L, Tan-Mullins M (2021) Pioneering a Green Belt and Road Initiative (BRI) alignment between China and other members: mapping BRI's sustainability plan. Blue-Green Syst 3(1):49–61
12. City Watch (2019) The sword points to the Southwest Economic Center! Don't underestimate this frontier city group. <剑指西南经济中心!千万别小看这个边疆城市群>. Available from: https://baijiahao.baidu.com/s?id=1630478488615381761&wfr=spider&for=pc
13. CXXXG (2017) Planning of central Yunnan urban agglomeration (2016–2049) <滇中城市群规划(2016–2049年)>. Available from: https://www.cxxxg.net/news-id-2747.html
14. Deng M, Li W, Hu Y (2016) Decomposing industrial energy-related CO_2 emissions in Yunnan province, China: switching to low-carbon economic growth. Energies 9(1):23
15. Dong E (2005) Regulations on the environmental management of tourism development in Yunnan Province, China. In: Bricker K (Comp. & ed) Proceedings of the 2004 Northeastern recreation research symposium: Gen. Tech. Rep. NE-326, pp 192–199
16. Fang C (2015) Important progress and future direction of studies on China's urban agglomerations. J Geog Sci 25(8):1003–1024
17. Fang C, Yu D (2020) Safeguard measures and suggestions for China's urban agglomeration construction. In: China's urban agglomerations. Springer, Singapore, pp 235–265
18. Hong J, Shen Q, Xue F (2016) A multi-regional structural path analysis of the energy supply chain in China's construction industry. Energy Policy 92:56–68
19. Jiang Y, Lin Y, Sun X (2012) Green development and ecological construction. In: China green development index report 2011. Springer, Berlin, pp 237–261
20. Kun W, Jinming C, Jianguang S, Rong G, Shixiang G (2021) Optimal allocation and integrated dispatch of regional water resources in Yunnan Province—problems and countermeasures. In: IOP Conference series: earth and environmental science, vol 831, No 1. IOP Publishing, p 012065
21. Liu J, Tian Y, Huang K, Yi T (2021) Spatial-temporal differentiation of the coupling coordinated development of regional energy-economy-ecology system: a case study of the Yangtze River economic belt. Ecol Ind 124:107394
22. Peng L, Deng W, Zhang H, Sun J, Xiong J (2019) Focus on economy or ecology? a three-dimensional trade-off based on ecological carrying capacity in southwest China. Nat Resour Model 32(2):e12201
23. Ptak T, Konrad V (2021) "Crossing the River by Feeling the Stones": how borders, energy development and ongoing experimentation shape the dynamic transformation of Yunnan Province. J BordLands Stud 36(5):765–789
24. Shi JZ (2015) Supernormal growth: China's economy 1979–2049. World Scientific, Singapore
25. Sun X, An H, Liu X (2018) Network analysis of Chinese provincial economies. Phys A 492:1168–1180

26. Tang W (2019) Research on the development of yunnan characteristic towns in the new era. In: The international conference on emerging researches in management, business, finance and economics (ERMBFE 2019). Available from: https://webofproceedings.org/proceedings_series/ECOM/ERMBFE%202019/ERMBFE182.pdf
27. Tian J, Zeng Y (2011) The management mechanism of education exchange and cooperation platform for Southeast and South Asia in Yunnan China. In: Advances in multimedia, software engineering and computing, vol 2. Springer, Berlin, pp 365–370
28. Wang R, Bai Y, Alatalo JM, Guo G, Yang Z, Yang Z, Yang W (2022) Impacts of urbanization at city cluster scale on ecosystem services along an urban-rural gradient: a case study of central Yunnan city cluster, China
29. Wu Q, Cheng J, Liu D, Han L, Yang Y (2015) Kunming: A regional international mega city in Southwest China. In: Urban development challenges, risks and resilience in Asian Mega cities. Springer, Tokyo, pp 323–347
30. Wu X, Wu J, Deng K, Chen H, Wang Y (2018) Study on industrial structure adjustment of fuxianhu waterside based on water environment carrying capacity. in the international conference on economics, politics and business management (ICEPBM 2018). Available from: https://webofproceedings.org/proceedings_series/ECOM/ICEPBM%202018/ICEPBM092.pdf
31. Xu M, Zhang K (2021) The value of ecological assets based on land cover/use pattern in Kunming, Yunnan Province. Research Square. Available from: https://assets.researchsquare.com/files/rs-1123701/v1_covered.pdf?c=1639435019
32. Yang R (2006) Yunnan's water resource development: problems and prospects. China Rep 42(1):25–39
33. Yi S (2021) Make coordinated efforts to advance the building of modern circulation system in the context of the new development pattern. China Financ Econ Rev 10(3):111–128
34. Yili Group (2022) China's first zero-carbon food factory: Yili Group's Yunnan facility certified as achieving carbon neutrality. A news report available from: https://www.prnewswire.com/in/news-releases/china-s-first-zero-carbon-food-factory-yili-group-s-yunnan-facility-certified-as-achieving-carbon-neutrality-804586942.html
35. Yunnan Government (2020) Notice of the People's Government of Yunnan Province on the planning and development of urban agglomeration in central Yunnan (2020). <云南省人民政府关于印发滇中城市群规划发展的通知> Available from: http://www.yn.gov.cn/zwgk/zcwj/zxwj/202008/t20200826_209715.html
36. Yunnan Urban and Rural Planning Design Institute (2017) Urban agglomeration plan of Central Yunnan. <滇中城市群规划(2016–2049年) >. Available from: https://www.ynghy.cn/h-nd-132.html
37. Zeng X, Liu E (2017) The studies of the cooperation model of Yunnan and neighboring countries in the restructuring of production capacity. In: 2nd international symposium on business corporation and development in South-East and South Asia under B&R initiative (ISBCD 2017). Atlantis Press, pp 166–171
38. Zhang Z, Wang B, Buyantuev A, He X, Gao W, Wang Y, Yang Z (2019) Urban agglomeration of Kunming and Yuxi cities in Yunnan, China: the relative importance of government policy drivers and environmental constraints. Landscape Ecol 34(3):663–679
39. Zong W, Cheng L, Xia N, Jiang P, Wei X, Zhang F, Jin B, Zhou J, Li M (2018) New technical framework for assessing the spatial pattern of land development in Yunnan Province, China: a "production-life-ecology" perspective. Habitat Int 80:28–40

Chapter 14
Carbon Neutrality and China's City Cluster Development: Future Directions

14.1 City Cluster Development at a Glance

As we have mapped in this book, China's city cluster development is a large-scale and long-term development plan. Its linkages with China's sustainable development plan and carbon neutrality plan are not coincidental. It is also part of China's new-type urbanisation process and high-quality urbanisation plan. The step-by-step process of developing several main regions helps the country to become more connected domestically and internationally. This approach, in a way, promotes urbanisation and innovation, which is the backbone of China's primary agenda to reach higher positions in the global stance. According to China [7] "*city clusters and its intended urbanization may help because urban dwellers with higher incomes generally consume more than rural dwellers. Clustering can also increase China's productivity and innovation by grouping more firms and increase the size of the labor market*". Thus, we see this long-term plan of regional development could help China in multiple ways, one of which is to accomplish its carbon neutrality pledge.

According to Cheshmehzangi and Chen [3, 4], there are several main challenges in meeting the sustainability plan and reaching carbon neutrality, which means there are still significant bottlenecks in China's decarbonisation process, industrial restructuring, and low-carbon transformations. Also, [22] highlight at least four overarching challenges, including (1) the continuously increasing energy consumption and CO_2 emissions, (2) a difficult process for China to reduce its carbon emission reduction through the ongoing transitional phases, (3) over-reliance on high-carbon fossil energy with high energy consumption and low energy utilisation efficiency, and (4) a relatively lower-level economic development in compared to other countries or regions (such as the US, the EU, and other developed nations), and with higher economic risks. Thus, we note the current decade until the end of the sustainable development goals (SDGs)' timeline is critical to China's plan to meet some of its key targets, such as the carbon peak targets in 2030.

Amid the COVID-19 pandemic crisis, China has opted to focus more on enhancing the transportation and communications links within the country. To some extent,

A. Cheshmehzangi and T. Tang, *China's City Cluster Development in the Race to Carbon Neutrality*, Urban Sustainability, https://doi.org/10.1007/978-981-19-7673-5_14

regional development was experimented with in different ways, providing opportunities to strengthen regional cooperation, provincial measures, local action plans, etc. On the contrary, we note the ongoing pandemic has also caused major disruptions to some of China's operations and development with the regional countries and neighbours. For instance, we can see severe impacts on China's ongoing projects with the ASEAN countries of Southeast Asia, which were also meant to speed up some of the internal projects and programmes in China (e.g., the major transportation link projects between Yunnan and the ASEAN region). In addition, some other programmes, such as the development of the carbon trading market [3] and carbon intensity reductions, are also affected severely. Thus, we expect to see some delays that may potentially be impacted by other externalities, such as lack of investment.

China's plan for carbon neutrality was somewhat inevitable as a response to climate change and its impacts on human and natural habitats. As highlighted throughout the book, China's city clusters have proposed to have their localised measures for achieving carbon neutrality. Each cluster has proposed its unique low-carbon transitions and decarbonisation processes. In essence, some of the decision-making and policymaking processes are aligned with regional capacity, the status of development, economy, infrastructural availability and provision, institutional support, and investment. Indubitably, some of the regions have had to start some of the initiatives much earlier, and some received (and continue to receive) higher-level attention for their ongoing and future development plans. Moreover, as China is shifting to mandate its dual circulation strategy, it should also aim to boost sustainable production, reduce consumption, increase efficiency across the board, and reduce CO_2 emissions at a gradual pace. Therefore, the introduction to the energy mix and gradual transitions are essential to China's step-by-step process toward decarbonisation, low carbon development, carbon intensity reduction, and the ultimate carbon neutrality goal.

Serving as a conclusion chapter, this chapter provides discussions based on what has been mapped in the book. From the case study examples of city clusters (Chaps. 3–13), we identify many similarities/commonalities and differences between regions. While similar challenges and opportunities exist between different city clusters, the governments and relevant stakeholders/actors must not promote one-fit-all solutions or strategies for all regions. We trust strategies and pathways to carbon neutrality should be contextualised, and each region should come up with its own direction, timeline, and plans. However, a careful monitory of the progress is essential to ensure higher-level targets and goals are achieved in a given timeline. In this chapter, we summarise China's future directions for China's city clusters under the influence of the carbon neutrality plan. To sum up our discussions, we provide nine potential directions before concluding the book.

14.2 From a Pledge to Reality: Can China Achieve Carbon Neutrality? and How?

After almost two years after the announcement of the pledge to carbon neutrality, China has invested more in its city cluster development plan. As the infrastructural development would eventually reduce and the city-level GDPs would slow down due to the real estate conditions, China would need to focus more on its urban–rural relations, industrial restructuring, and high-quality urbanisation. But can the most populous country achieve carbon neutrality by 2060? Could this pledge become real? And will it be a process that may be disrupted during its timeline? While the book only mapped case study examples, we still highlighted how the carbon neutrality plan is ongoing and developing at the regional levels. The city clusters are where policy development, interventions, and implementations are expected to happen. Thus we can say: yes—the pledge can become a reality if all steps are carefully taken and if other externalities do not affect the whole process of almost four decades; and yes, the process could take a different turn at any moment, should it be for good or worse, and could incorporate new technologies, innovation, and new policy development.

China's pledge to carbon neutrality has initiated three primary directions. The first is China's domestic development plans and consolidated methods of meeting the targets at multiple scales. The second is the growing domestic and international competitiveness, which has initiated a new race in the carbon neutrality arena. For instance, several other countries have also announced their carbon neutrality plans after China's pledge. This publicising outcome was somehow boosted by China's earlier announcement in autumn 2020. And the third is opportunities for multilateral, regional, sub-regional, and bilateral cooperation that were not robust enough before. The potential global climate policy is more visible than before, showing China is a leading player in low-carbon and sustainability areas.

In this so-called climate diplomacy landscape, China's attempts are more ambitious than just the earlier experimentations and short-lived interventions. China's learning phase is completed already, which has led to speeding the process of the sustainable development plan in many ways. Many experimental projects and programmes were initiated in these one or two decades, showing that China has explored different modes of development, strategies, and directions. The city cluster development plan has become an ongoing—and long-term—plan, where each region is empowered, supported, and guided in different ways. Despite differences in how solutions should be strategised at the regional level, we note that there are also common future directions that are highlighted in the following section.

14.3 Future Directions for China's City Clusters

Here, we summarise nine potential directions and—to some extent—opportunities for China's city clusters. The below specification is just a summary of what we have discovered from mapping 11 of the 19 city clusters and investigating them in accordance with China's sustainable development plan and carbon neutrality goals.

14.3.1 Energy Transformation Path and Industrial Restructuring

Clean energy transition, sustainable urban development, and investment priorities are considered key areas for China's energy transformation [11], particularly in the city cluster context. These concerning areas would lead to improvements in air quality, carbon intensity levels, and meeting the carbon peak targets of 2030 [3]. Also, air quality co-benefits should be highlighted in this transition [12]. In addition, industrial restructuring should be promoted through different mechanisms, such as investment in building low-carbon cities and developing low-carbon or zero-carbon projects. Other aspects to consider are associated with "*reducing industrial coal consumption through restructuring and technological progress*" [11]. Thus, the importance of regional-level interventions rather than national-level strategies for industrial restructuring.

14.3.2 From Institutional and Infrastructural Development to Social and Policy Development

In the process of combatting climate change and its impacts, institutional and infrastructural development are deemed critical [6]. Most of China's super and medium-level city clusters have already achieved a relatively suitable infrastructure to support their carbon neutrality plans. However, there is still a need for developing the energy grid, renewable energy capacity, delivery and utilisation, and integrated systems for both urban and rural areas [17]. Nevertheless, we anticipate the shift to be gradually towards social and policy development, including lifestyle changes, development of legislation, incentives, and support for low carbon transitions and decarbonisation projects. In this transition, the current status of infrastructures alone cannot achieve the carbon neutrality targets. Therefore, the need for smart systems, integrated planning [24], and technology integration is essential. According to [16], the governments and local actors should increase "*investment in low-carbon projects and improve infrastructure construction*", which then requires longer-term planning through mechanisms like the circular economy strategy.

14.3.3 A Holistic Process of Policy Development, Monitory, Control, and Review

At the regional level, provincial, municipal, and local governments should develop or adopt contextualised policies related to their specific needs and conditions. Thus, a holistic policy development process must be completed with careful monitoring, control, and regular reviews. To do so, local governments should follow specific guidelines and develop their own policies. A monitory process at the municipal and provincial levels is then needed to ensure certain aspects are checked and cross-checked adequately. For instance, carbon concentration, socio-economic development, and target measurements are critical in this process [23]. Moreover, "*monitoring can help with the dynamic measurement of the changes in the whole life cycle of carbon dioxide*" (ibid). Such considerations should also highlight geographical and ecological environmental changes that may differ from region to region. Thus, once again, a regional approach is vital so that progress is not only monitored or controlled but is also reviewed and reflected on as prompt as possible.

14.3.4 Development of New Standards, Higher-Level Benchmarks, and Contextualised Green Development

The city clusters also need to focus on developing new green-, eco-, and low-carbon standards to ensure benchmarks are reviewed and adjusted when required [5]. The so-called contextualised green development could help enhance the current experimental projects at smaller scales and towards scaling up processes. This process should be deliberately contextualised to ensure green development is aligned with social lifestyles, cultural values, economic status, etc. Such development could help augment environmental protection and ecological preservation strategies and programmes at the regional level. Reducing carbon footprint and its pressure help reduce carbon emissions and balance economic development with environmental protection [9]. For instance, research and practice in carbon offset [18] could promote the importance of environmental protection. Other aspects, such as carbon emission efficiency [20] need to be the highlight of the development process for standards, benchmarks, and green strategies.

14.3.5 From Rapid Urbanisation to High-Quality Urbanisation

The idea of high-quality urbanisation is very much linked with high-quality economic development [19]. Embedded in the 13th and 14th Five-Year Plans (FYPs), China has gradually shifted away from its rapid urbanisation process and toward high-quality urbanisation. This approach reverses some of the earlier trends of poor construction, inconsiderate development, low-quality infrastructure and services, etc. According to [8], the relationship between urbanisation and green technology innovation is crucial, particularly when the intention is to promote "*the development of the technology economy and accelerate the flow and exchange of technology and knowledge*". Therefore, approaches to high-quality urbanisation should include technological upgrading and transformational opportunities for new developments, urban and rural renovations, infrastructural development, smart applications, etc. We argue in favour of high-quality urbanisation, which requires high-level measures for carbon footprint reductions, carbon emission reduction, carbon intensity reduction, and many more such measures aligned with sustainable development directions.

14.3.6 Environmental Protection Plans, Ecological Restoration, and Low Impact Development

At the city cluster level, factors like environmental protection, ecological protection, and low-impact development seem to become more apparent than at the smaller scales of cities and districts. The links between 'urban and rural', 'urban and urban', and 'rural and rural' become much better for larger-scale environmental protection plans. Compared to most developed countries, China's ecological protection is still behind and requires further consideration of low-impact development, ecological capacity, carbon footprint, etc. [1]. As we are currently in the transition from low carbon to carbon neutrality [21], we expect to see impacts on "*the economies of energy-producing and heavy industrial cities, as well as the more developed high-tech and service-oriented cities that depend on them, so radical industrial restructuring or large-scale shifts of the country's energy mix is unlikely to happen anytime soon*". In light of this transitional phase, we see the importance of low-impact development emerging faster than before, allowing for transformational development trends or patterns that may consider more of the environmental protection and ecological restoration agenda.

14.3.7 Planning Strategies at Multiple Spatial Levels for Integrated Innovation of Green and Low-Carbon Technologies

The main direction is to consider methods of carbon emission control that could update the industrial clusters. In doing so, China is expected to consider an integrative path for its city cluster development [21]. As [13], "*energy consumption has to become carbon neutral as soon as possible and in an integrated approach that includes familiar goals for achieving carbon neutrality*". Thus, planning strategies must take a multi-level and multi-objective approach [15], which could help promote green development. Other associated aspects that need to be considered are related to methods of integrated innovation that could include green and low-carbon technologies, not only at smaller scales of buildings and neighbourhoods, but in urban systems and infrastructure of larger scales, such as the urban, rural, township, and regional scales. As proposed by [16], a key direction is to consider multi-energy integration of science and technology innovation, a shift that is expected for mature cities, developing the industrial clusters, and leading enterprises to be low carbon and then carbon neutral. An example of this is the integration of renewable energy and a better energy mix into multiple sectors, industrial clusters, and the energy grid, and enhancing urban–rural relations.

14.3.8 Consolidated Governance for Circular Economy and High-Quality Economic Development

A study by [2] indicates the spatial importance and impacts of the circular economy on carbon intensity. Therefore, we can argue in favour of the city cluster development, where a circular economy can be successfully implemented on a large scale. Led by high-quality economic development, the focus could become more comprehensive, adapting methods of sustainable development through better institutional support and governance. Fang et al. [10] highlighted that the performance of urban circular economy promotion depends on carbon footprint measurements. The same applies at the regional level, where we can reduce carbon footprints through better urban–rural development mechanisms. At the regional level, we can have healthier mobility between the urban and non-urban areas, better support from rural to urban areas (and vice versa), better renewable energy capacity, etc. For instance, we can have a higher number of recycling centers, multiple ways of generating energy from renewable means, and integrate a smart and comprehensive grid system. Lastly, for carbon neutrality, high-quality economic development is essential [19] mainly because it could create a healthy relationship between multiple sectors and systems in both the urban and non-urban areas. This approach is essential for China's high-quality urbanisation and, ultimately the carbon neutrality plan.

14.3.9 Technological Innovation, Integration, and Beyond Mere Experimentation

The "*promotion of low-carbon technology of the entire chain*" is important in boosting science and technology innovation [16]. For instance, in the cases of new energy public transportation, China can focus more on regional links and systems [14] as well as smaller-scale interventions. However, technological innovation is not enough, and integration into the existing and new urban areas is crucial. At the city cluster level, carbon technological progress is essential, which has a much more substantial impact on emission reduction [9].

For instance, [8] explore the impact of green technology innovation on increasing carbon emission efficiency, which is vital for regional sustainable development approaches. The example of industrial clustering, as mapped in many of the studied city clusters in this book, shows the role of technological innovation and integration. For instance, larger-scale economic circles, industrial parks, and other related industrial clustering are ways forward beyond just experimentation projects. We believe China has already developed and accomplished many experimental projects that should now be moving towards scaling up and larger-scale implementation. If to be strategised adequately, technologies will become critical drivers for the enhancement of energy-related and emission-related factors; thus, the push for their integration into existing infrastructure, services, and communication links is ever essential.

14.4 In the Race to 2030 Carbon Peak and 2060 Carbon Neutral Plan

The race to carbon neutrality is inevitable, as it is our utmost human nature to be competitive and eager to work with indicators, targets, and goals. China's 2020 pledge to the 2060 carbon neutrality plan opened up at least five types of competition (or perhaps competitiveness) at multiple scales. At the global level/scale, the race to achieve carbon neutrality has intensified since China joined the competition (as it seems). It has also created a sort of movement between countries through multi-lateral relations to ensure countries of specific regions could join hands and come up with larger-scale interventions. Some examples include multi-lateral programmes in the EU, South Asia, the Middle East (particularly the Arab nations alongside the southern coasts of the Persian Gulf), Central Asia, South America, the ASEAN region, North America, and Oceania. The second type is competition, which is boosted through bilateral relations. After China's announcement of its carbon neutrality plan, countries, particularly in eastern and south-eastern Asia (and some in south Asia and the African continent), are focused more on working collaboratively and cooperatively with China through country-to-country relations and mechanisms. These cooperation are still weak and require better coordination. Despite the COVID-19 impacts on mobility, trade, and business, several plans are ongoing to boost sustainable marketing, trade, and development via bilateral relations. At the national level, there is competition between the regions (or clusters), finding ways through new policy development and strategies to meet the 2030 carbon peak targets and the 2060 goals. This is the level where 'city clusters' exit and play a significant part in developing or adopting new pathways aligned with the national-level policy and guidelines.

At the regional level, there is also a sort of competition between larger cities. The race is comprised of multiple competitions, e.g., to become more prominent in their respective regions, more advanced and modernised, gain more financial support from national-level and provincial-level governments, and bid for national-level pilot city programmes. In regions with no large or mega-cities, the city cluster development would boost the race between cities of medium-to-large scale to decide which one becomes the most prominent city of the cluster. The same competition exists in industrial clusters and economic circles. For instance, in the Chengdu-Chongqing economic circle or city cluster, discussions on changing the political boundaries of the two provinces are back on the table again. More recently, Chongqing has proposed to add Guang'an (located in the far east of the Sichuan province and neighbouring Chongqing) to its territories, which is still an ongoing debate between the Chongqing and Sichuan governments. Lastly, there is competition between districts of cities, townships, and surrounding areas. The carbon neutrality plan has opened many taps of new financial support and mechanisms, allowing for smaller-scale projects, city-city programmes, and urban–rural development opportunities. One of the recent examples is the recently funded project of multi-million dollars that focuses on carbon neutrality plans for three cities of Chengdu, Chongqing, and Ningbo. Also, at the district level or in the surrounding areas of cities, local governments bid for funding for their new pilot projects aligned with the overarching carbon neutrality plan, financing their industrial restructuring projects and boosting their economic growth and development. The race to carbon neutrality at all these five scales could become more vivid and attention-grabbing to study in the near future.

Nonetheless, we note that race is not necessarily a fair one within the context of China. There are apparent differences between the clusters of different categories of super, medium, and small. Thus, as we noted throughout the book, contextualised evaluation, planning, interventions, and directions are necessary. Here, we conclude the book with the hope that we can revisit the status of China's carbon neutrality plan after the year 2030 and see how the progress has been over the years in this critical decade. We hope the steps are taken cautiously, inclusively, and fairly to allow opportunities for all and boost the economies of each region while considering the importance of environmental protection, social infrastructure development, education, cultural values, ecological restoration, provision of infrastructure and services, and equality in prosperity distribution. China's path to carbon neutrality is unique; thus, it makes sense to develop new ideas and innovative directions. The contextualised opportunities should become enabler drivers to resolve problems of the city clusters and address them in the possible comprehensive approach. As we highlighted, the race is ongoing, and we have around 38 more years until the proposed carbon neutrality plan is reached.

As much as they are the potential to succeed, there are possibilities to fail, and China must take its steps with a clear long-term plan and open vision to avoid any disruptions, open up for cooperation, and welcome innovation, upgrading, and scaling up opportunities. We also hope to remain influential actors from the research and academic side to ensure playing our part in helping the following steps, learn from the past and elsewhere, and find innovative ways that could genuinely reverse some of the recent unsustainable trends. The war against climate change is not over; we are losing it in many ways. But we require a robust willingness, support, and reinforcement to ensure we can make a genuine difference with positive impacts on the planet. This is the war started and fueled by ourselves, and it has to teach us lessons before we come up with solutions. If we opt not to change our mindsets and paths, we may continue repeating our previous mistakes. And if we opt to be inclined toward genuine paradigm changes, then we may just be able to think not only outside the box but also to think without the presence of the box. For us, the next four decades are super exciting, and we trust many countries will join the race and achieve their goals more cooperatively. At any time from now until 2060, an utterly game-changing technology could change everything, but we hope that we rely more on the basic foundations of humanity to ensure sustainable development is not far from reach. The next step must be to create healthy living environments for all and not only for us humans. It should include new values, indicators, targets, and paradigms, and we hope carbon neutrality is achieved through new thinking and innovation and not the business-as-usual directions. We trust that China's city cluster development plan will be assessed, monitored, and reviewed carefully to ensure the goal is unified and enthused for all, i.e., to reach the carbon neutrality plan.

References

1. Chen B, Xu C, Wu Y, Li Z, Song M, Shen Z (2022) Spatiotemporal carbon emissions across the spectrum of Chinese cities: insights from socioeconomic characteristics and ecological capacity. J Environ Manage 306:114510
2. Chen P (2022) The spatial impacts of the circular economy on carbon intensity-new evidence from the super-efficient SBM-DEA model. Energy Environ 0958305X221125125
3. Cheshmehzangi A, Chen H (2021a) China's Sustainability transitions: low carbon and climate-resilient plan for carbon neutral 2060. Springer Nature
4. Cheshmehzangi A, Chen H (2021b) Decarbonised race and new destination in China. In: China's sustainability transitions. Springer, Singapore, pp 53–69
5. Cheshmehzangi A, Chen H (2021c) Key suggestions and steps ahead for China's carbon neutrality plan. In: China's sustainability transitions. Springer, Singapore, pp 189–206
6. Cheshmehzangi A, Dawodu A (2019) Introduction: climate change and cities—perspectives, planning and people. In: Sustainable urban development in the age of climate change. Palgrave Macmillan, Singapore, pp 1–22
7. China Briefing (2018) China's city clusters plan to transform into 19 super regions. Available from: https://www.china-briefing.com/news/chinas-city-clusters-plan-to-transform-into-19-super-regions/
8. Dong F, Zhu J, Li Y, Chen Y, Gao Y, Hu M, Qin C, Sun J (2022) How green technology innovation affects carbon emission efficiency: evidence from developed countries proposing carbon neutrality targets. Environ Sci Pollut Res 29(24):35780–35799
9. Fan W, Huang S, Yu Y, Xu Y, Cheng S (2022) Decomposition and decoupling analysis of carbon footprint pressure in China's cities. J Clean Prod 133792
10. Fang K, Dong L, Ren J, Zhang Q, Han L, Fu H (2017) Carbon footprints of urban transition: tracking circular economy promotions in Guiyang, China. Ecol Model 365:30–44
11. Hepburn C, Qi Y, Stern N, Ward B, Xie C, Zenghelis D (2021) Towards carbon neutrality and China's 14th five-year plan: clean energy transition, sustainable urban development, and investment priorities. Environ Sci Ecotechnology 8:100130

12. Liu J, Zhou W, Yang J, Ren H, Zakeri B, Tong D, Guo Y, Klimont Z, Zhu T, Tang X, Yi H (2022) Importing or self-dependent: energy transition in Beijing towards carbon neutrality and the air pollution reduction co-benefits. Clim Change 173(3):1–24
13. Ramanathan V, Allison J, Auffhammer M, Auston D, Barnosky AD, Chiang L, Collins WD, Davis SJ, Forman F, Hecht SB, Kammen DM (2016) Bending the curve: ten scalable solutions for carbon neutrality and climate stability. Collabra 2(1)
14. Shi T, Si S, Chan J, Zhou L (2021) The carbon emission reduction effect of technological innovation on the transportation industry and its spatial heterogeneity: evidence from China. Atmosphere 12(9):1169
15. Sun J, Cheshmehzangi A, Wang S (2020) Green infrastructure practice and a sustainability key performance indicators framework for neighbourhood-level construction of sponge city programme. J Environ Prot 11(2):82–109
16. Sun X, Zhang H, Ahmad M, Xue C (2022) Analysis of influencing factors of carbon emissions in resource-based cities in the Yellow River basin under carbon neutrality target. Environ Sci Pollut Res 29(16):23847–23860
17. Tan-Mullins M, Cheshmehzangi A, Chien S, Xie L (2017) Smart-eco cities in China: trends and city profiles 2016. Exeter: University of Exeter (SMART-ECO Cities Project)
18. Wei J, Zhao K, Zhang L, Yang R, Wang M (2021) Exploring development and evolutionary trends in carbon offset research: a bibliometric perspective. Environ Sci Pollut Res 28(15):18850–18869
19. Xiao R, Pan L, Xiao H, Xiao H, Zhu Z (2022) Research of intelligent logistics and high-quality economy development for Yangtze River cold chain shipping based on carbon neutrality. J Mar Sci Eng 10(8):1029
20. Zhang H, Feng C, Zhou X (2022) Going carbon-neutral in China: does the low-carbon city pilot policy improve carbon emission efficiency? Sustain Prod Consum 33:312–329
21. Zhang Z, Hu G, Mu X, Kong L (2022) From low carbon to carbon neutrality: a bibliometric analysis of the status, evolution and development trend. J Environ Manage 322:116087
22. Zhao X, Ma X, Chen B, Shang Y, Song M (2022) Challenges toward carbon neutrality in China: strategies and countermeasures. In: Resource, conservation and recycling, vol 176, p 105959
23. Zhou C, Zhang R, Loginova J, Sharma V, Zhang Z, Qian Z (2022) Institutional logic of carbon neutrality policies in China: what can we learn? Energies 15(12):4391
24. Zhou S, Li W, Lu Z, Lu Z (2022) A technical framework for integrating carbon emission peaking factors into the industrial green transformation planning of a city cluster in China. J Clean Prod 344:131091

Index

A. Cheshmehzangi and T. Tang, *China's City Cluster Development in the Race to Carbon Neutrality*, Urban Sustainability, https://doi.org/10.1007/978-981-19-7673-5